The Urban Book Series

Igor Vojnovic, Department of Geography, Michigan State University, East Lansing, MI, USA

Claudia van der Laag, Oslo, Norway

Qunshan Zhao, School of Social and Political Sciences, University of Glasgow, Glasgow, UK

The Urban Book Series is a resource for urban studies and geography research worldwide. It provides a unique and innovative resource for the latest developments in the field, nurturing a comprehensive and encompassing publication venue for urban studies, urban geography, planning and regional development.

The series publishes peer-reviewed volumes related to urbanization, sustainability, urban environments, sustainable urbanism, governance, globalization, urban and sustainable development, spatial and area studies, urban management, transport systems, urban infrastructure, urban dynamics, green cities and urban landscapes. It also invites research which documents urbanization processes and urban dynamics on a national, regional and local level, welcoming case studies, as well as comparative and applied research.

The series will appeal to urbanists, geographers, planners, engineers, architects, policy makers, and to all of those interested in a wide-ranging overview of contemporary urban studies and innovations in the field. It accepts monographs, edited volumes and textbooks.

Indexed by Scopus.

Qinyi Zhang

The Elemental Metropolis

The Past and Future of the Extended Urbanity
in the Yangtze River Delta, China

 Springer

Qinyi Zhang
Studio Paola Viganò
Antwerp, Belgium

ISSN 2365-757X   ISSN 2365-7588 (electronic)
The Urban Book Series
ISBN 978-3-031-36408-2   ISBN 978-3-031-36409-9 (eBook)
https://doi.org/10.1007/978-3-031-36409-9

This Springer imprint is published by the registered company Springer Nature Switzerland AG
The registered company address is: Gewerbestrasse 11, 6330 Cham, Switzerland

Foreword

The Elemental and Horizontal Metropolis

The Elemental Metropolis is a three-faced book. First, it investigates a segment of Chinese urbanization through a deconstructive approach that reveals the physical features of the territory through the mapping and the redesign of its fundamental elements. Special attention is paid to long-term aspects as well as to more recent landscape transformations. A close reading reveals the "material conditions" (the nature of the soil, the type of agriculture, the rental aspects, the different types of property, the economies…) that constitute the socio-ecological space of the Yangtze River Delta and the Tangqi region.

Second, the book develops the hypothesis that the current process of metropolization is still growing fast in China and comes on top of the rural rationalities of water, soil, and agriculture. Contemporary land use is a hybridity in which the different urban–rural characteristics intermingle and create a new type of space. The literature about this hybridization of metropolitan and rural conditions has changed a great deal in the last decades. This book adds some very interesting perspectives through an elementary deconstruction of the metropolis. The present research supports and enables the understanding of the processual adaptations and transformations of the territory to maintain its productivity, safety, accessibility, etc., according to different rationalistic thought processes designed for the maintenance and inhabitability of urban–rural–metropolitan China.

Third, the book reconstructs what has been deconstructed and creates a vision for a "*Horizontal Metropolis*". The oxymoron combining horizontal with metropolis is intended to open a fundamental debate about the role of space, spatiality, and spatial design in reconfiguring social, economic, and ecological relations (horizontal vs. vertical) and rights (the right to the city, as well as the right to stay put, the right to have a certain freedom of choice, the right to access and to emancipation). We are facing increasing marginalization of regions and territories everywhere on the planet, in favor of the accumulation and concentration of wealth and its related economies in very limited and attractive places. The cost of such concentration is a reversion to and exploitation of servants in large parts of the world. The extension of the metropolis invading nearby rural areas results in a tremendous loss of fertile soils, biodiversity, water management, food production and rural cultures, exacerbating the risks factors inherent in densely populated areas.

The *Horizontal Metropolis* paradigm acts as a counter project against this concentration–marginalization process. This book investigates this process through a research-by-design approach that recognizes the capacity of design to produce an original knowledge, while referring to a genealogy of ideas and authors whose network connects a wide range of contributions to the specificity of Chinese socio-political history.

The three faces contain the three-fold ambition of the book, bringing them together in a clear, lucid, and intelligent way, fed by the intellectual context in which these initial hypotheses have been originally traced and developed. Qinyi Zhang was involved in the International Consultation for the Great Paris vision proposals (2009) as a student of the European Master in Urbanism program. It has given him strong foundational skills that are reinforced by his work with the team reflecting on a vision for Brussels 2040 as a *Horizontal Metropolis*. It was during that process that Qinyi Zhang decided to question China's urbanization and to examine the metropolitan condition of Southwest China with a clear ambition to develop a proposal for its future. Mapping and design can become very explicit and extremely sophisticated activities. Starting from the conceptual tools used by the *Elementary City* and *Horizontal Metropolis* research, Qinyi Zhang tackled the process with great clarity and vision. The book charts the gradual deconstruction of the ancient and modern Chinese landscapes, while interrogating their capacity to maintain an emancipatory role, in which premodern, modern, and contemporary economies, infrastructures, family and social structures co-exist.

Beyond the traditional idea of the metropolis, this book outlines an alternative spatial and socio-ecological path and will stimulate discussion about this topic.

Milan, Italy   Paola Viganò

Acknowledgements

I would like to express my sincere gratitude to Prof. Bernardo Secchi and Prof. Paola Viganò. I came from China to Europe more than ten years ago, to study and practice architecture and urbanism. In most of this long period of studying and practicing, I am grateful that I am always guided by them. In such a complex and challenging profession, their tremendous knowledge, infinite curiosity, and extreme passion for design and research have been enlightening for me. This work of Ph.D., supervised by Prof. Viganò, is the most recent part of this guidance, and I cannot imagine it without her.

I would like to thank Prof. David Grahame Shane, the copromotor of my thesis, for the many constructive discussions through talks and emails. His great knowledge of China and urbanism has been incredibly inspiring.

I would like to thank Prof. Michiel Dehaene (UGent, Belgium) and Prof. Stephen Cairns (ETHZ, Switzerland) as external referees of my thesis, for their insightful and critical comments, which have been a great influence on my work. I would like to thank Prof. Maria Chiara Tosi and Prof. Stefano Munarin (IUAV) for their critics and encouragement.

I would like to thank Prof. Terry McGee and Prof. John Friedmann for the fruitful discussions. I would like to thank all the members of Lab-U in EPFL, and Andrea Palmioli from HKU, for multiple constructive exchanges. Especially I would like to thank Prof. Wang Zhu 王竹, Prof. Hua Chen 华晨, and Post-doc researcher Qian Zhenlan 钱振澜 from Zhejiang University for their critics and advice.

I would like to thank Alexander Wandl from TU Delft for his advice on the writing, and Ekaterina Andrusenko for her knowledge of Russian and Soviet urbanism.

I would like to thank my father Zhang Yuhan 章煜汉 for his tremendous help and local knowledge. The fieldwork in Tangqi could not be done without him and his social network. I would like to thank Kristien Wynants for her hard and professional work on the revision of my work. And finally, I would like to thank my wife Dai Min 戴敏 for supporting me tremulously throughout the writing of this thesis and my life in general.

Contents

Chapter 1
Introduction

Abstract The first part of this chapter explains the notion behind the title. The title "The Elemental Metropolis", which put elemental and metropolis together, introduces a seemly unusual way to look at the contemporary metropolitan urban condition. It proposes a hypothesis that a metropolis, with all its complexity, can be constructed and transformed through its basic elements. This hypothesis is then tested on a specific territory—the Yangtze River Delta in China. It is an untraditional "metropolis" that is undeniable a metropolis in population density, infrastructure, economy, and many other aspects, but also a radical mixture of rural and urban in terms of activities and space, therefore a *desakota* (The definition of the *desakota* by McGee is the regions "of an intense mixture of agricultural and nonagricultural activities that often stretch along corridors between large city cores...were previously characterized by dense populations engaged in agriculture, generally but not exclusively dominated by wet-rice" (McGee, In The extended metropolis: settlement transition in Asia. University of Hawaii Press, Honolulu, (1991): 10).). The chapter introduction draws a path through the concept of the metropolis, the *desakota*, and the notion of the elements. The second part of the chapter explains the main structure of the research and a brief introduction to each chapter of the book.

Keywords Metropolis · *desakota* · Elements · Structure

1.1 The Metropolis, the *desakota*, and the Elements

Yangtze River Delta (Fig. 1.1), the 358,000 km² alluvial plain in east China, is highly urbanized. More than 60% of its 235 million population lives in 41 cities, including three cities with more than 10 million inhabitants (Shanghai, Suzhou, and Hangzhou). Its population density (656 hab./km²) is comparable to some of the most famous metropolitan areas in the world, such as Île-de-France (1021 hab./km²) and Randstad (739 hab./km²), but on a much vaster area (approximately 30 times of Île-de-France and 40 times of Randstad). It is a vast territory with metropolitan conditions.

© The Author(s), under exclusive license to Springer Nature Switzerland AG 2023
Q. Zhang, *The Elemental Metropolis*, The Urban Book Series,
https://doi.org/10.1007/978-3-031-36409-9_1

|0 | | | | 250km|

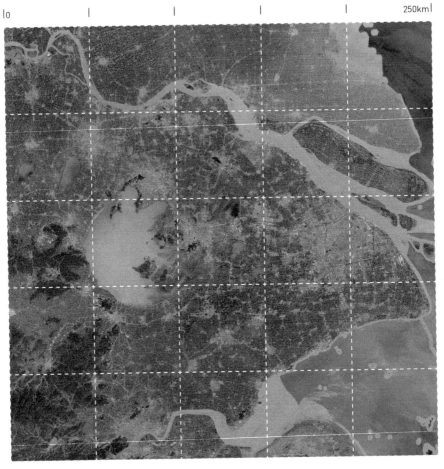

Fig. 1.1 Satellite photo of the Yangtze River Delta in the 250 × 250 km frame, 2014. *Source* Google earth. Elaborated by author

The image of such a metropolis is often symbolized by the skyline of Shanghai, the mega port of Ningbo, the high-speed trains, and, more and more, the preserved or restored historical villages in the countryside. But at the same time, the vast scale of such a metropolis is often neglected. What is the vast space in between the cities, the ports, and the few historical villages made of?

Today, a 159 km train journey from Shanghai to Hangzhou, two major cities respectively east and southwest of the Yangtze River Delta, takes 47 min. A passenger looking out of the window, first sees the Shanghai periphery, an extremely busy place, full of viaducts, large box-sized industrial buildings with blue roofs, and high-density residential areas, occasionally interrupted by a plot of agricultural land, a river, and a wood. After about 6 min, large open fields become more frequent. Before long, the passenger realizes that the openness of the field is a preemptive illusion. When looking more closely, the field contains a great number of houses, small industries (again, mostly with blue roofs), woods, winter gardens, fishponds, billboards, electrical towers, viaducts, rivers, and various other objects. Two or three towns or small cities pass in front of the window for just a couple of seconds. After 24 min, the train stops at a brand-new station in the middle of a field, with a couple of towers on the horizon. In the next 16 min, the same view continues, and the passenger cannot figure out where he/she is until the train enters the periphery of Hangzhou (Fig. 1.2). What is this space?

A group of Western and Chinese scholars wrote about this space, and some defined it as an object. Wittfogel categorizes the Yangtze River Basin as the typical Asiatic and hydraulic society, heavily reliant on the building of large-scale irrigation works (Wittfogel 1957). Fei's case study of Kaixiangong Village in the southern part of the Yangtze River Delta depicts the delicate irrigation system of the paddy field, the dense and diffuse urbanization and craft workshops, and the emergence of village industry (Fei 1939). Huang points out the saturation of land productivity and the need to relocate surplus labor as the main challenges of the Yangtze River Delta (Huang 1990), which is echoed by the later work of Wen (2009). Friedmann's writing about coastal China also highlights the high rural density and surplus labor and suggests diffuse urbanization as an "endogenous" process (Friedmann 2005). In terms of the form of urbanization, Gottmann listed the "urban constellation around Shanghai" as one of the six global megalopolises (Gottmann 1976), and his notion of corridor implicitly influenced later concepts such as the Metropolitan Interlocking Regions Nanjing-Shanghai-Hangzhou in Yangtze River Delta by Zhou (1991).

Within this body of knowledge, the concept of the *desakota* by McGee (1991) as regions "of an intense mixture of agricultural and nonagricultural activities that often stretch along corridors between large city cores…were previously characterized by dense populations engaged in agriculture, generally but not exclusively dominated by wet-rice" (1991: 10) clarifies the essence of this space. More than the other scholars, he first roughly identifies the location of the *desakota* regions, the Yangtze River Delta as one of them, and then constructs a model to conceptually show the spatial relation of those regions with the major cities and hinterlands. From the 1980s to the 2010s, the model of the *desakota* has evolved, but in general, the *desakota* is defined around the large urban cores, but with a peri-urban space between the *desakota* and the urban cores. If the peri-urban is directly impacted by the expansion of the urban core, the *desakota* is the space where the exogenous drivers from the city, nation, and the world, and the endogenous drivers of surplus labor, the productivity of land, local knowledge of entrepreneurs, etc., interact with each other.[1] In other words, there is a higher level of independence within the *desakota* region: if in the peri-urban, people mostly commute daily to the urban cores, people in the *desakota* work in the in-situ industries and the field. If the increasing density of the peri-urban areas is a result of migration from the countryside and urban inhabitants pushed out of the city, the

---

[1] See McGee's description of "exogenous" and "endogenous" drivers, and "above ground knowledge" and "ground knowledge" (McGee 2016). The diffuse urbanization in China as an "endogenous" process is also well stated in Friedmann's work on China (Friedmann 2005).

Fig. 1.2 Screenshots of the video were taken from a high-speed train from Shanghai to Hangzhou. The screenshots are taken every 30 s. The video shows a continuous space with dispersed and repeated elements: housing, factories, infrastructure, farmland, vegetable fields, water ponds, billboards, etc. Made by the author

high density in the *desakota* is due to historical in-situ urbanization linked to the specificity of its agricultural production.

The model of the *desakota* was constructed for a few Asian regions in general. By closely examining the Yangtze River Delta via the definition of the *desakota*, one can find it on a much broader scale. It is a space that fills the entire territory except for the urban cores and the remote villages, covering about 2/3 of the delta. The form of this *desakota* goes beyond the area around the urban cores or along the corridors in

Fig. 1.2   (continued)

between them and becomes a vast and universal layer. This layer is introduced in this thesis as the "third" space. The "third-ness" is found in many aspects different from the urban cores and the remote villages: its diffuse and dense form of urbanization, its isotropic and fine road network with hardly any major infrastructure corridors, its immense number of dispersed small industries, but without major industrial estates, etc. The term "third space" in this thesis is not intended to be a replacement for the *desakota* model but an addition to it. If the *desakota* defines its area in terms of its population, actors, social-economical drivers, flows, administration, and mix of activities, the "third space" attempts to complement it with the materiality of the territory—the physical dimension of the *desakota*.

By studying the physical dimensions of the *desakota*, this thesis attempts to answer a very basic question: what constitutes the *desakota* in the Yangtze River Delta?

This question is particularly relevant in today's urbanization in China. After a long tradition of diffuse urbanization, from the 1990s urbanization entered the city-centered stage. Today, the city-centered urbanization model, still a classic way of planning in China, has hit its limits: the over-crowed cities, traffic congestion, the environmental crisis, the shortage of resources, etc. A new paradigm is needed. At the same time, the city-centered approach, in which the demographic, economic, and social issues are dealt with relatively, neglects the environment of the non-cities; the full and rich spatial content of the *desakota* is not recognized.[2] This absence of recognition of the physicality of the *desakota* limits the possibility to establish a new paradigm.

The main methodology used to approach this question is a close examination of the spatial elements, through which the *desakota* is made and functions: the water, the trees, the houses, the road, the industry, the facilities, etc. The notion of elements is strongly presented in Secchi's "Progetto di Suolo", Viganò's "La città elementare", Gregotti's "The Form of the Territory", Venturi's "Learning from Las Vegas", Wright's "The Living City" and "The Broadacre City", Branzi's "The No-stop City", Koolhaas' "The Generic City", and many others. The main reference referred to here is the work of "La città elementare" by Viganò, in which the study of the elements is considered a gesture of "step-back": retreating to the most basic and detailed study of the elements of the fragmented reality, a minimum but fundamental certainty can be generated prior to interpretation and allows maximum space to construct new possibilities. To put it simply, my thesis attempts to continue the exercise of Viganò of the city of Prato in a broader and different context: the metropolis of the Yangtze River Delta.

The *desakota* as an interpretation like every other interpretation implies a project—a project of the future of the territorial metropolitan area that is deeply rooted in its past. It is not enough to describe the content of each element, or the physical dimensions of the *desakota*, through its past and status quo; the description has to be complemented by the dimension of its potential future—its capacity to be sustained, used, and transformed into a new society is part of its nature. Therefore, this thesis goes further, to imagine the elements in a possible project for the future of the *desakota*. Since this thesis is part of a broader research program on the Horizontal Metropolis, directed by Prof. Viganò in EPFL in Lausanne, the trajectory from elements to a project is exercised via a common protocol shared by studies on other territories globally. The main part of the conceptualization is done via a physical model, which has the same dimensions and scale as the "Broadacre City" model by F. L. Wright.

That explains the title: The elemental metropolis: The past and future of the Extended Urbanity in the Yangtze River Delta, China.

---

[2] See the recognition of dispersed territories as a natural and spatial capital (Viganò et al. 2016).

1.2   The Structure of the Research

The main body of the book is structured into five chapters.

Chapter 2 Yangtze River Delta: the discovery of a "third" space

The first part, on the *desakota* in the Yangtze River Delta: a process of urbanization, presents the consistent process of dispersed urbanization with the mixture of industry and agriculture via a brief history of the different stages of urbanization in China. It explains the inner mechanism through which a delicate balance of the climate and geography of the territory, the type of agriculture, and the density of the settlements in the agriculture economy stage. It describes the Maoist setup of the communes as the first step toward rural industrialization and in-situ urbanization, which was subsequently unchained after the Economic Reform of the 1980s, and finally, the take-over of the city-centered urbanization in the 1990s. A brief analysis of the limits of the current city-centered urbanization process is made. The second part of the chapter is dedicated to the "third" space concept or the physical and spatial appearance of the *desakota*. We make the comparison to some western city-territory paradigms and give an overview of the latest research and policies relevant to the third space. The third part of this chapter deals with the notion of element(s) and explains the methodology of this thesis which is based on elementalism. I reflect briefly on the term elementalism as applied by a group of architects including Gregotti, Venturi, Secchi, and Viganò. Furthermore, our methodology analyzes the elements starting from their layers and eventually moves on to the systems, based on the work of Viganò's *La città elementare*.

Chapter 3 Elements: A study of Tangqi

The first part of this chapter elaborates on the reason we chose the town of Tangqi, located in the southwestern along the Yangtze River Delta as the subject for our case study. My intensive fieldwork in Tangqi uncovers several elements which are further explained in the second part. A micro-story of these elements and their history, social-economical background, functionality, and a description of their physical performance accompanied by photos and interviews complete our analysis. Thus, the rationality, the transformation, and the challenges related to each element are laid bare. The third part deals with the urban–rural split by the elements and reflects on the articulation of the elements, the notion of public space in a rural and urban setting, and a comparison between the two.

Chapter 4 From elements to layers

To tackle the complexity of the territory, we try to connect the local scale in which the elements are found and the territorial scale, in which the repetition of the elements could create a critical mass. The author created a territorial frame of 50 × 50 km according to a series of conditions for a typical third space, presented by a collection of maps. Several layers are presented and the selection of the elements for the layers is based on the elements from the Tangqi case study described in chapter two. Due to a great lack of data, I put great effort into collecting historical maps and GIS data and produced a detailed tracing of elements on a territorial scale based on Google Earth satellite images. This means that the mapping of some of the layers—for instance, the trees, the industrial production space, and the Socialist New Villages—has never been carried out before and is truly original, especially on such a large scale. The final part of this chapter includes a reading of the territorial structure via the layers.

Chapter 5 Imagining a utopia

This chapter has two parts. The first part, "Atlas of a utopia", is dedicated to constructing an atlas of works by a small group of thinkers and designers, including Kropotkin, Frank L. Wright, the Soviet disurbanists, the Japanese Shirakaba-ha group, and Mao, not as the work of a historian, but as an exercise to find a common social and spatial agenda among them toward a unity of the urban and the rural. The atlas is used as a reference utopia for the *desakota* of the Yangtze River Delta. The second part, "Imagining a utopia", is an exercise not in providing concrete solutions but in imagining specific qualities of life in a society where the urban–rural divide is eliminated by transforming the elements. This is accomplished using a set of systems, in which the social and spatial agenda is imagined within the possibilities offered by the specificity of the space and conditions of the *desakota*.

Chapter 6 Conclusion

The conclusion offers reflections on the *desakota* as a specific physical environment, and its limit to the interaction between the exogenous and endogenous forces, resisting the latter while overwhelmed by the former. It is followed by a description of the "Elementary Metropolis" as a type of work. The "Elementary Metropolis" presents a type of research with a sequence of steps, a complete trajectory based on the elementalism through which the physicality of a metropolis can be explored.

The thesis ends with a bibliography and an appendix. The appendix consists of the translation of a selection of works by the Soviet disurbanists, which still have great relevance today and are extremely inspiring. It is the first time that they have been translated into English.

The research for this thesis was conducted as part of the research on the Horizontal Metropolis directed by Prof. Viganò in EPFL in Lausanne, as a part of a broader tradition of research on regions with dispersed conditions, with closely interlinked, copenetrating rural/urban realms, communication, transport, and economic systems. A common protocol embodied in the research of Horizontal Metropolis is implemented here, including micro-story construction, thematic mappings, comparison, etc., which makes it possible to connect this work with the research on other contexts around the world. This work also benefits from the long experience of the author as a student of Bernardo Secchi and Paola Viganò and working with Studio Bernardo Secchi Paola Viganò and later Studio Paola Viganò on different projects in Europe, especially territorial visions such as Grand Paris, Brussels 2040, and Great Moscow.

References

Fei X (1939) Peasant life in China. E. P. Dutton Company, New York

Friedmann J (2005) China's urban transition. University of Minnesota Press, Minneapolis

Gottmann J (1976) Megalopolitan systems around the world. Ekistics 243(February):109–113

Huang P (1990) The peasant family and rural development in the Yangzi delta, 1350–1988. Stanford University Press, Stanford, California

McGee TG (1991) The emergence of *desakota* regions in Asia: expanding a hypothesis. In: Ginsburg N, Koppel B, McGee TG (eds) The extended metropolis: settlement transition in Asia. University of Hawaii Press, Honolulu

McGee TG (2016) Città Diffusa and Kotadesasasi: comparing diffuse urbanization in Europe and Asia. A project in the making. International roundtable conference: territories of metropolis: compactness, dispersion, ecology: comparative perspectives between Asia and Europe. 5 to 7 April 2016 Shanghai, China

Viganò P, Secchi, B, Fabian L (Eds) (2016) Water and asphalt. The project of isotropy. Park Books, Zurich

Wen T (2009) 三农问题与制度变迁. China Economic Publishing House, Beijing

Wittfogel KA (1957) Oriental despotism. A comparative study of total power. Yale University Press, New Haven

Zhou Y (1991) The metropolitan interlocking region in China: a preliminary hypothesis. In: Ginsburg N, Koppel B, McGee TG (eds) The extended metropolis: settlement transition in Asia. University of Hawaii Press, Honolulu, pp 89–111

Chapter 2
The Yangtze River Delta
and the Discovery of a Third Space

Abstract The first part, on the *desakota* in the Yangtze River Delta: a process of urbanization, presents the consistent process of dispersed urbanization with the mixture of industry and agriculture via a brief history of the different stages of urbanization in China. It explains the inner mechanism through which a delicate balance of the climate and geography of the territory, the type of agriculture, and the density of the settlements in the agriculture economy stage, the Maoist setup of the communes as the first step toward rural industrialization and in-situ urbanization. This was subsequently unchained in the third stage after the Economic Reform of the 1980s and afterward, and finally, city-centered urbanization took over in the 1990s. A brief analysis of the limits of the current city-centered urbanization process is made. The second part of the chapter is dedicated to the "third" space concept or the physical and spatial appearance of the *desakota* (Fig. 2.1). We make the comparison to some western city-territory paradigms and give an overview of the latest research and policies relevant to the third space. The third part of this chapter deals with the notion of element(s) and explains the methodology of this thesis which is based on elementalism. I reflect briefly on the term elementalism as applied by a group of architects including Gregotti, Venturi, Secchi, and Viganò. Furthermore, our methodology analyzes the elements starting from their layers and eventually moves on to the systems, based on the work of Viganò's *La città elementare*.

Keywords *desakota* · Process of urbanization · In-situ urbanization · City-centered urbanization · The third space · Elementalism

2.1 *Desakota* in Yangtze River Delta: A Process of Urbanization

This part attempts to situate the study on the Yangtze River Delta in the broader context of China's urbanization. It starts with a description of an often neglected but still strongly present Chinese agricultural economy, which culminates in the Yangtze River Delta. This economy contributed significantly to the development of the territory, and the combination of small agriculture and domestic industry within

© The Author(s), under exclusive license to Springer Nature Switzerland AG 2023
Q. Zhang, *The Elemental Metropolis*, The Urban Book Series,
https://doi.org/10.1007/978-3-031-36409-9_2

it, reveals the characteristics of what McGee defines as *desakota*: regions with an intense mixture of agricultural and non-agricultural activities that often stretch along corridors between large city cores (McGee 1991: 7). This particular type of agricultural economy provides a basic but structural layer for the present and future urbanization process. It also generates an "involution" process in which productivity and income per unit of labor shrink as more labor is invested, an intensification without development, which has been the fundamental challenge of unbalanced urban–rural development.

The description is followed by an explanation of the three stages of urbanization in China, starting with a decentralized, in-situ urbanization process to concentrated, city-centered urbanization. This thesis has no intention of adding a new reading of the urbanization history in China but simply tries to clarify the relation between the different industrialization models with this underlying layer, the relation between those models, and the "involution" process. Finally, the author presents a critical reading of the limitations of the current concentration model in the Yangtze River Delta and the ongoing search for a new paradigm of territorial development. This search motivates the re-examination of the overlooked agricultural and territorial layer as the starting point of the thesis.

2.1.1   The Agricultural Economy

In "*Oriental Despotism: A Comparative Study of Total Power*" (Wittfogel 1957), Karl Wittfogel published a highly criticized definition of "Asian" society and an equally strongly criticized reflection on Russia and communist China. He constructs a link between a specific type of agriculture—one that relies on large-scale irrigation or oriental agriculture—with a specific organization of society—the hydraulic society, and a specific path to industrialization. Wittfogel quoted from Marx's articles in the New York Daily Tribune in 1853, "climate and territorial conditions" made "artificial irrigation by canals and waterworks the basis of Oriental agriculture." (Wittfogel 1957: 373), and "the economic structure of Chinese society" is "dependent upon a combination of small agriculture and domestic industry" (Wittfogel 1957: 374). The Asiatic state is formed on top of that structure. Moreover, Wittfogel and Marx realize that the age-long perpetuation of the Asiatic state is the result of a type of space: the "dispersed" condition of the "oriental people" and their agglomeration in "self-supporting" villages (combining small agriculture and domestic handicraft) (Wittfogel 1957: 374).

The climate, the territory, the irrigated agriculture with its vast hydraulic infrastructure, the dispersed population, self-supporting villages, and the combination of small agriculture and domestic handicraft/industry display a new paradigm, clearly distinct from the classic western society. This specificity or "orientality" has been noticed by a series of more recent scholars, such as Friedmann and McGee. It not only describes a different historical condition but also indicates a possible alternative path to the future. This path differs from the one described by the unilinealists of the nineteenth century and in the case of Wittfogel, a path different from the "classical" development sequence that, according to Friedmann, occurred in Western Europe (Friedmann 1981: 246). The path also moves away from the "Western paradigm of the urban transition, which draws its rationale from the historical experience of urbanization such as took place in Western Europe and North America in the nineteenth and twentieth centuries" (McGee 1991: 5) for McGee. If this paradigm is to remain valid today, a shift toward a new form of urbanization may have to impose itself. This model (path) differs from the city-making movement featuring high-density residential development and industrial estates.

This particular "oriental" paradigm is well explained in the works of Fei (1939, 1948), Huang (1990), and Wen (2009).[1] The most significant characteristic of this type of agricultural economy is the co-existence of high population density, highly productive land, and high pressure from surplus labor. From the Song Dynasty (960–1279) to the New China, the arable land per capita has been low and shrinking despite the growth in arable land, due to the dense and growing population.[2] In the mid-1990s, the arable land per capita in China was 0.08 ha/capita, 42% of that of India, and 33% of the global average. On the other hand, the arable land has been highly productive: its productivity rose 3.5 times from 1949 to 2005. In 1991, the land productivity index (1422) reached almost 3 times the world average (515), 3.5 times that of the USA (410), and close to that of Japan (1744). However, productivity per unit of labor is extremely low. In 1991, it (422) was 39% of the world average (1080), 8% of that of the USA (5156), and 86% of that of India (493)—too many laborers on too little land.[3] The surplus laborers have sought alternative sources of income—in the historical silk crafts and today's factories. From 1996 to 2005, the percentage of income from the non-agricultural sector rose from less than 25 to 35 for each rural worker.[4]

This paradigm never saw a fundamental reformation in many Chinese regions, and certainly not in the Yangtze River Delta. The Wittfogelian concept of hydraulic society, which draws a causal relationship between the type of agriculture and the form of society in China, is generally true and could be applied to the Yangtze River Delta. The agricultural economy that allows each peasant to own a small plot of the arable field while at the same time preventing the accumulation of wealth is an

---

[1] A relation exists between the work of Wittfogel, Fei, Huang, and Wen. Wittfogel and Fei met in Columbia University in 1943, and both lectured in Harvard University. In Wittfogel's reflection on China, Fei's work is often quoted. In his work Wen mentions the debate on the "Asiatic society" in the 30 s within the communist party of China. The unilinealists won and the supporters of the Asiatic society theory were defined as Trotskyists, for whom Wen expresses sympathy and appreciation (Wen 2009, page 11). In his work, Huang criticized the work of Wittfogel, and revises Wittfogel's alternative path with additional diversity and specificities of different regions of China. (Huang 1990: 148).

[2] Source: Wen (2009). pp 44.

[3] Source: Li (1996).

[4] Source: rural household survey 2005, National Bureau of Statistics of the People's Republic of China.

accurate observation, the Yangtze River Delta being its perfect example.[5] Statistics make this even more clear: as arable land (for all types of agricultural production) per capita in China is 0.12 ha, it is only 0.065 in the Yangtze River Delta; the productivity per hectare is much higher in the Yangtze River Delta than the national average (99,162 yuan/ha versus 64,901 yuan/ha); the productivity per person is much lower in the Yangtze River Delta than the national average (6445.5 yuan/capita versus 7788.1 yuan/capita).[6] However, this geographical determinism has been criticized and revised by other scholars. In the *Peasant Family and Rural Development in the Yangzi Delta, 1350–1988* by Huang, the specificity of the Yangtze River Delta is proven by a comparative study of various rural societies in both northern and southern China: compared to authority constituted by the central empire and local gentries (in the form of official hereditary committees) as Wittfogel explains, the distribution of political power in the villages of the Yangtze River Delta is much more diffuse and horizontal. The villages are subdivided into different locations, and collective work is often organized within each subsettlement. Authority is mostly determined by individual events instead of being permanent. Moreover, highly developed commercial activity has not led to the collapse of rural social groups as Marxists predicted. One obvious reason is simply the existence of the extensive water network that surrounds the land plots (Huang 1990: 149)—the type of territory of the Yangtze River Delta generates variation within the universal hydraulic society. At the same time, due to the Chinese regime and its manner of managing urbanization, few scholars have been able to construct a discourse on the Yangtze River Delta alone without involving all of China, especially the unprecedented social-economic reforms that took place after 1949, as its background. The main references are made on those two levels, and many works cover both: some references are made on a national Chinese level, including Friedmann (2005), McGee (1991), McGee et al. (2007), Wen (2009), Wittfogel (1957), Zhou (1991), etc.; the others are made on the Yangtze River Delta level including Fei's (1939) case study of Kaixiangong Village, Huang (1990), McGee (1991), McGee et al. (2007) case study on the Shanghai-Nanjing *desakota* area and Kunshan, Zhou's (1991) case study on the Nanjing–Shanghai–Hangzhou metropolitan interlocking region, etc.

The Yangtze River Delta can be used as a representative case to explain in detail the relationship between a territory and its agricultural economy. The climate makes for a 300-day-a-year agricultural season and precipitation of around 1200 mm per year. The delta slightly inclines from the north to the south and from the east to the west. The periphery of the delta is 3–5 m above sea level, while most of the center, where Tai Lake is located, is lower than 3 m above sea level—the whole delta is thus a bowl with a few higher hills scattered on it. The topography leads to the production of rice in most of the delta, and cotton on the higher periphery. Rice is raised in paddy fields, a specific water field. Fei explains the mechanism operating between this type of agriculture and dense settlements, with a specific case

---

[5] When Wittfogel claims it as a general condition in China, he refers to the work of Fei who has made a case study of the Yangtze River Delta. (Wittfogel 1957: 80).

[6] Source: Statistical Yearbook 2016, National Bureau of Statistics of People's Republic of China, elaborated by author.

Fig. 2.1 A Chinese *desakota*: a continues "carpet" of rural settlement where agricultural and non-agricultural activities are highly mixed. Drone photo of the town of Tangqi, Hangzhou, China, an ordinary town in the Yangtze River delta. Taken by the author

study on Kaixiangong village, a typical village on the southern bank of Tai Lake. The productivity of the land is on the one hand among the highest in China, but on the other hand, is also saturated. The concentration of labor and plots of land will not enhance the effectiveness. Each laborer was capable of cultivating a field of 0.4 ha (Fei 1939: 152), which directly defined a precise population density of around 800/km². To manage the water, the field has to be organized in *yu* (圩, a piece of land surrounded by water) subdivided into *jin* (塝, smaller plots divided by dikes) (Fig. 2.2). The irrigation and drainage of each *jin* call for collective work among families. It assembles an incredibly dense network of villages within 20 min walking distance of each other (Figs. 2.1 and 2.3).[7,8]

The small plots of land owned by each family produce barely enough food for domestic consumption. The rice field season runs from June to the beginning of

---

[7] McGee also notes the specificity of paddy field. As he quotes Spate and Learmonth (Spate and Learmonth 1967: 202), "Paddy has developed a strikingly similar landscape, broadly similar from the Ganga to the Yangtze… but no other way of life… has led to the evolution of a cultural system so stable and permanent as that associated with great paddy-plains of Monsoon Asia." (McGee 1991: 3).

[8] The research of Horizontal Metropolis by Viganò in EPFL in Lausanne, targeting on the regions with closely interlinked, copenetrating rural/urban realms, communication, transport and economic systems, produces interesting comparison between those regions. This thesis is part of the research. See also other searches and PhD thesis under the frame of Horizontal Metropolis, and other participated studies, working with the Eastern and Western contexts. They include the works of Martina Barcelloni Corte on Thailand and Andrea Palmioli on China.

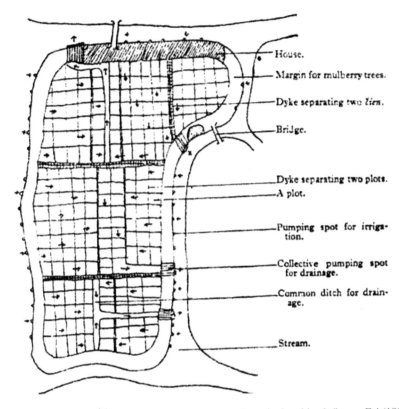

House.

Margin for mulberry trees.

Dyke separating two *lien*.

Bridge.

Dyke separating two plots.

A plot.

Pumping spot for irrigation.

Collective pumping spot for drainage.

Common ditch for drainage.

Stream.

Fig. 2.2 Structure of *yu* 圩: a unit of water management in agricultural land. *Source* Fei (1939b)

December, with a vacation from July to September. Women and children are not involved in intensive fieldwork. The redundant laborer is relocated to textile production—especially silk production and inter-village trade for extra income to cover necessary family expenses. The bowl-shaped topography and rich precipitation in the delta forced the inhabitants to construct one of the densest river, canal, and lake networks in the world—an extremely well-developed water transportation system that enables the villages to trade and exchange their products. At the junctions of the waterways, an immense number of towns have emerged. Although the Shanghai-Hangzhou railway exists since 1909, a huge number of villages and towns rely on this diffuse water network. Fei observed that the first village ship factory was founded in the 1930s in Kaixiangong village, located distant from the railway. John Friedmann compares the development of cities in Flanders in the middle of the eleventh century, as described by Fernand Braudel, to that in the lower course of the Yangtze River in the same period as a "flourishing, rapidly urbanizing economy". Since then, the life of a small farmer in the Yangtze River Delta has been a mixture of agricultural and non-agricultural activities. This tradition of diversified production and expanded

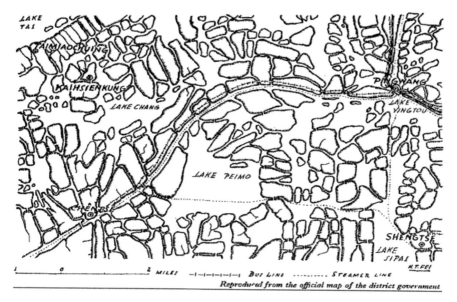

Fig. 2.3 Surroundings of the Kaixiangong Village, a landscape of the yu (圩). *Source* Fei (1939b)

markets on the fertile rice fields "is feeding contemporary rural industrialization in coastal China." (Friedmann 2005: 43).

As a result, a delicate but stable balance has been established between population density, the productivity of rice agriculture, and business based on the water network. The saturated land productivity does not demand investment in more labor, meaning that redundant laborers have no choice but to constantly struggle to find other sources of income to survive. Different from the definition of intensification (more labor, same productivity per unit of labor per day) and development (more labor, higher productivity per unit of labor per day), Geertz and Huang define it as involution: the productivity and income per unit of labor per day shrink as more labor is invested. The struggle is not always successful. Fei recorded in the 1930s that large-scale implementation of advanced machinery in Japan led to a collapse of the textile industry in the Delta and a great number of laborers returning to the agricultural sector. After 1978, this "endogenous" motivation found its outlet in the process of industrialization and urbanization. Relocating the enormous labor surplus in agriculture became the main challenge in rural China.

2.1.2 The Three Stages of Urbanization

As mentioned above, the combination of the climate, the territory, the irrigation agriculture with its vast hydraulic infrastructure, the dispersed population, the self-supporting villages, and the mixture of small agriculture and domestic handicraft/

industry forms the basis, or in Marx's words, the structure, for future industrialization and urbanization. After the establishment of the New China, stages of urbanization can be pinpointed: the Maoist era, rural industrialization and in-situ urbanization, and city-centered urbanization. Through these three stages, an often-overlooked continuous and dispersed type of urbanization is found in the countryside next to the eye-catching urbanization in Chinese cities.

2.1.3   The Maoist Era: The City of Production

Radical land reform took place at the beginning of the 1950s. In 1953, every peasant in China owned land, and there was no relocation of the surplus agricultural population. This exacerbated the dispersal of the laborers. As the industrial sector was booming in the cities during the first five-year plan (1953–1957), around 20 million migrants from the countryside boosted the urban population from 10.1% to 19.7% from 1949 to 1960. From 1958 to 1978, a strict household registration—the *hukou*—was implemented, which divided the population into the non-agricultural and the agricultural and prevented migration toward the urban area.

Mao perceived and was excited about the "endogenous"[9] power in the countryside. In his report "*About the Rural Collectivization*" (1955), he predicted, "a national socialistic reform will arrive soon in the countryside, which is inevitable". He did not doubt that an approach had to be taken different from the Stalinist one that over-exploits the countryside for the development of industry and cities. In 1961, he stated the need to avoid the redundant rural population rushing into the city, and therefore the necessity to improve the standard in the life standard in countryside to be "the same as the one in the city, or better". For him, such an improvement can be realized through commune construction, and each commune transforms peasants to work in situ by its economic center and industry (Mao 1998: 197).[10]

---

[9] The word "endogenous" was used by Friedmann to describe the motivation of seeking alternatives within the rural area as the fundamental reason for China's urban development: "instead of globalization…an evolutionary process that is driven from within, as a form of endogenous development." (Friedmann 2005: XVI).

[10] Mao's idea of urbanizing the countryside coincides with Okhitovich and Ginzburg's idea of "disurbanization" through the competition entry for Green City in Moscow (1930) and Magnitogorsk (1930), where radical elimination of the difference between urban and rural areas was deeply rooted in communist thinking by Lenin, Marx and Engels. The Maoist communes, especially their light industry focusing on local conditions and targeting domestic needs, the internal exchange of merchandise, and their scale in size and population, strongly influenced the Agropolitan District that Friedmann proposed in the 1970s, as a "city-in-the-fields" (Friedmann and Douglass 1978). It also coincides with much utopian thought and practice in the countries of the former Soviet Union and Eastern Europe, including under-urbanization, less urbanism and a relatively uniform urban space, as noted by McGee et al. (2007: 11).
    See Chap. 4.

According to Meisner,[11] this direction was implemented during the Cultural Revolution, despite many constraints. "The establishment of industries in non-urban areas under the collective management of communes and brigades (townships)" was encouraged. (Chan 1994: 81). By 1978, the rural industries that were collectively managed employed 28% of China's total industrial laborers. Although urbanization seems to have declined in terms of numbers, the growing rural industries employed more than 20 million rural laborers and established many communes as centers of knowledge and technology. Meisner considers the rural industrialization plan as the most successful part of Mao's vision of a self-sufficient countryside and rural–urban equivalence.[12]

In the Maoist era, the role of a city and the countryside was production rather than consumption. Collective units (*danwei*) were organized in cities to maximize output and manage distribution. The communes and the brigades are the production units in the rural areas. In each commune and brigade, one could find the complete package of collective housing, public facilities, rice fields, orchards, and woods; a collective and modern version of the self-supporting countryside of the past. The family industry was kept parallel to collective agriculture.[13] The industries, infrastructure, medical services, schools, and other facilities were developed rapidly in the countryside—an industrialization layer was added to the agricultural economic structure recognized by Wittfogel and Marx. Together they created the basis for future urbanization and industrialization in China. Today, many traces from that time can be recognized from space: roads lined with tall trees, primary schools, ports, and bridges built in the countryside, and more recently privatized factories built in the commune era. The commune as a rural production unit was set up at the edge or even inside the city to abolish the border between urban and rural—one can imagine a near invisible transition from the city of Beijing to its surroundings, or even to its furthest corners: simply an endless territory of units and communes.[14]

It is interesting to note that the scale of a commune, although varying dramatically, often follows the existing pattern of villages and towns in the Yangtze River Delta. The ancient structure of settlements was not fundamentally changed but integrated into the in-situ urbanization process.

---

[11] See "Social Results of the Cultural Revolution" (Meisner 1977: 352).

[12] Scholars have various opinions on the reason for this relocation of the urban population. According to Sit (2015), the relocation was completely political; according to Wen, the concentration of capital and exclusion of labor led by prioritizing the development of heavy industry was the main motivation. (Wen 2009: 178).

[13] See Huang's case study on Huangyang Brigades. (Huang 1990: 210).

[14] The idea of a city of production and the exchange of land between urban and rural has a strong influence on Friedmann's agropolitan development concept: "In addition, agropolitan districts may also be formed within the perimeter of large and growing cities (metrocenters). Just as agropolitan development attempts to bring the city to the countryside, so the countryside can be brought to the city ("fields in the city"). In other words, a modified agropolitan model can be introduced to help restructure the form of existing large cities by opening the built-up areas of the city to fields and urban farming". (Friedmann and Douglass 1978: 185).

Nevertheless, the tendency toward involution in the countryside—a process ongoing since 1350 in the Yangtze River Delta—was not reversed. The explosive increase in labor due both to the participation of women in the production process and rapid population growth after 1949 exacerbated the pressure on arable land. Collective agriculture was simply the enlargement of family agriculture. In general, the absence of a consistent urban development guideline led to disorganized urbanization (Li 2008: 27).

2.1.4 Rural Industrialization and In-Situ Urbanization

The economic reform begun in 1978 opened the lid on rural–industrialization and in-situ urbanization, which started to reverse the involution process. This may be a result of China relaxing its rules on two issues: one is the abolition of the collective system of agricultural production and the adoption of a household responsibility system (家庭联产承包责任制)[15]; the other is the relaxation of state control on rural economy, which caused a boom in town and village enterprises with numbers shooting up from 164,000 enterprises in 1978 to 1,740,000 in 1989.[16] The two measures brought extremely dispersed industrialization: by the middle of the 1990s, a city like Kunshan had rural industries in almost every village.[17]

Industrialization brought a significant rise in the income of the rural population. The gap between urban and rural incomes was reduced: the average income in urban areas was 2.6 times that in rural areas in 1978, and 1.5 by the end of the 1980s. It also significantly changed the composition of the rural economy. From 1978 to 1986, the share of agriculture in the total output of the rural area dropped from 60 to 40%, the share of forestry, animal husbandry, crafts, and fisheries remained 20%, and the share of industry increased from 20 to 40%.[18,19] Although the growth of rural industry slowed after the 1990s, it still plays a significant role in the economy of China today. In 2006, 34.6% of agricultural workers' income was from rural enterprises. In 2007,

---

[15] The household responsibility system was a practice in China first adopted in agriculture in 1979 and later extended to other sectors of the economy, by which local managers (the household) are held responsible for the profits and losses of an enterprise. In the traditional Maoist organization of the rural economy and that of other collective programs, farmers were given a quota of goods to produce by the government and received compensation for meeting these quotas. Going beyond the quota rarely produced a sizeable economic reward. In the early 1980s, smallholders were given drastically reduced quotas. What food they grew beyond the quota was sold on the free market at unregulated prices.

[16] Source: Statistical Yearbook 1982–1990, National Bureau of Statistics of People's Republic of China; Statistical Bulletin on Social Services Development in 2013, Ministry of Civil Affairs of People's Republic of China.

[17] See McGee's case study on Kunshan. (McGee et al. 2007: 146).

[18] Source: Statistical Yearbook 1982–1990, National Bureau of Statistics of People's Republic of China; Statistical Bulletin on Social Services Development in 2013, Ministry of Civil Affairs of People's Republic of China.

[19] See Huang's detailed case study of Huangyangqiao Commune in the Yangtze River Delta (Huang 1990).

|0 | 1km|

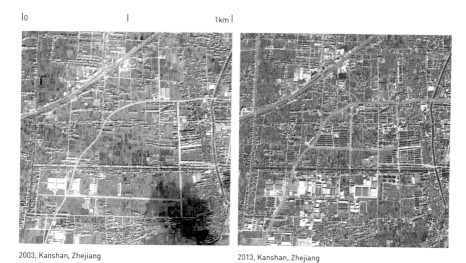

2003, Kanshan, Zhejiang   2013, Kanshan, Zhejiang

Fig. 2.4 Still ongoing process of rural industrialization and in-situ urbanization in the Yangtze River Delta. Kanshan, Zhejiang. *Source* Google earth. Elaborated by the author

rural enterprises employed more than 150 million laborers, 29.3% of which are rural laborers; the added value of the rural industry is 46.5% of the total industry (Feng 2008).

Parallel to rural industrialization, urban development policy in the cities and towns consistently stimulated a dispersion of small cities and towns, instead of a concentration of big cities. From "limit the scale of large cities and develop a great number of small cities and towns" in 1978, to "limit the scale of large cities, reasonably develop medium cities, and actively develop small cities" in 1980, and to "strictly limit the scale of large cities, reasonably develop medium and small cities" in 1990, the intention of limiting large-scale cities was leveraged while the development of small cities and towns was promoted, which echoes the "limit the development of large cities, create and expand small cities" principle of the 1st Five-Year Plan (1952–1957) of the Maoist era. During this period (the 1980s to 1990s), small cities and towns started to grow rapidly in the Pearl River Delta, the Yangtze River Delta, the coastal provinces, the Beijing–Tianjin–Hebei area, the three provinces in northeast China, and Sichuan—the *desakota* areas described by McGee, the interlocking metropolitan areas defined by Zhou and the multi-centric urban field by Friedmann (2005: 50).

The town and village enterprises were absorbing huge numbers of local surplus laborers from the agriculture sector—most of the employees of those enterprises were from nearby and did not move to the town and cities. Therefore, a great number of half-agricultural and half-industrial villages emerged, as a process of in-situ urbanization.[20] This in-situ urbanization is, instead of a consequence of the development of cities, a parallel development to it (Fig. 2.4). Different from the next urbanization

---

[20] Huang even calls this in-situ urbanization "industrialization without urbanization" (Huang 1990: 292).

stage, it can still be read as part of the continuity of the Maoist era: the initial town and village enterprises were transformed from the commune and brigade industries; the initial capital invested in the rural industry was the collectively owned surplus income of the commune; the laborers were mostly local inhabitants who still maintained their agricultural activities, rather than migrants. The industrialization layer on top of the agricultural economy was dramatically intensified. Therefore, Friedmann states China as "the only one" whose rural industrialization being "truly transformative". (Friedmann 2005: 39).

The recognition of this type of urbanization and its uniqueness is shared between both Western and Chinese scholars. Sit claims this pervasion of small cities and towns as a "bottom-up" movement (Sit 2015); McGee describes it as a "two-track" process,[21] which is "almost unique in world history: one in which the problems of surging rural–urban migration, inadequate urban infrastructure, and unemployment that characterized so many developing countries in the early phases of their urban transition, might be avoided." (McGee et al. 2007: 3); based on careful investigation of China in the early 1990s, Guldin, describes this process as a rising "urban lifestyle" in the countryside, which suggests the elimination of the urban–rural division. (Guldin 1997).

The process of involution was reversed and true development—in terms of individual productivity and income—happened in rural areas. This development is generally attributed to rural industrialization rather than to the growth in the agricultural sector—the productivity of agriculture was not improved after the economic reform (Huang 1990: 286). However, due to the increase in income that boosted demand for vegetables, fish, fruit, etc., the cash crops were planted more extensively. The traditional paddy fields were partly replaced by orchards, nursery gardens, fishponds, tobacco, saccharin, etc. A new countryside landscape began to emerge as a side effect of industrialization.[22]

2.1.5 The City-Centered Urbanization

From the 1990s, urban development policy turned to a city-centered one, especially a concentration of urban and economic activity in the urban cores. Vast development zones and industrial parks (*kaifaqu*) were built. From 1984 to 1991, there were only 14 national *kaifaqu* in China; from 1992 to 2016, 209 new national *kaifaqu* and 146 national High Technology Industrial Development Zones were established. Unlike the previous stage, this new industrialization is largely dependent on global

---

[21] The "two-track" process refers to both domestic investment in most parts of China and foreign investment in some southern area in China that took place in the in-situ urbanization process. (McGee et al. 2007: 3).

[22] See a case study of Tangqi on orchards and fishponds in Chap. 3.

investment and targeting the global market.[23] Large infrastructure, usually in the form of a regular grid, is often imposed on the existing pattern of the agricultural field. Vast concentrated industrial campuses, along with multi-level residential neighborhoods, are expanding in rural areas adjacent to central cities and towns. The laborers often come from less developed parts of China and inhabit the villages surrounding the industrial estates. They represent a modern and rational organization of industrial production, rely extensively on global capital and markets, and have caused a gigantic wave of migration of rural laborers from all parts of China rushing into the coastal urban area. The number of migrants increased dramatically, from 21.3 million in 1990 to 245 million in 2013,[24] which coincides with the development of the *kaifaqu*. The consumption of land is even more dramatic: from 2000 to 2011, the urbanized area in China grew by 76.4%, much higher than the growth of the urban population (50.5%); in the same period, the rural population shrunk by 133 million, but rural settlements expanded by 2.04 million ha.[25] In the Yangtze River Delta, the average yearly rate of expansion of the urban area from 1994 to 2012 was 7.8%, more than double the yearly growth of the urban population (3.7%).[26]

A new guideline on urbanization, and consequently a new system and hierarchy of cities, has emerged. The previous guideline, limitation of the scale of large cities, simply failed and was officially abolished[27] (Li 2008: 31). From the 10th Five-Year Plan (2000–2005) to the most recent 13th Five-Year Plan, it has been replaced by "a coordinated growth of large, medium and small cities". Li pointed out that in recent years (up to 2007), the number of cities has remained stable, while the number of large cities, especially the extra-large cities, has grown fast, and the number of medium and small cities has declined. The number of large cities (more than 1 million inhabitants) increased from 29 in 1978 to 40 in 2000, and rocketed to 133 in 2013; meanwhile, the number of towns increased from 2173 in 1978 to 20,312 in 2000 and has stagnated since then and reduced slightly to 20,117 in 2013.[28]

The Yangtze River Delta is one of the main districts where this new city-centered model took hold, presented clearly by the expansion of built area around the major urban cores after 2000s (Fig. 2.5). The fast development of the concentrated industrial and residential area of Kunshan is a well-studied case of such a model (Fig. 2.6). As McGee recognizes in his study in Kunshan, "tendencies for urban concentration" are exhibited in coastal China, in which the inhabitants in expanded built-up areas

---

[23] On the other hand, it also cooperates to some extent with the intense layer of rural industries. See in Chap. 3 a case study on industrial production in Tangqi.

[24] Source: National population census of 1982, 1990, 2000, 2010, National Bureau of Statistics of People's Republic of China; Report of national statistics on domestic economy and social development, 2011, 2012, 2013, National Bureau of Statistics of People's Republic of China.

[25] Source: The National Plan of New Type of Urbanization (2014–2020), Ministry of Human Resources and Social Security of the People's Republic of China (MOHRSS), 2014.

[26] Source: Zhou et al. (2016).

[27] In 2007, the new *China's law of city and town planning* deleted the previous planning guideline.

[28] Source: Statistical Yearbook from 1978 to 2005, National Bureau of Statistics of People's Republic of China; Statistical Bulletin on Social Services Development in 2013, Ministry of Civil Affairs of People's Republic of China.

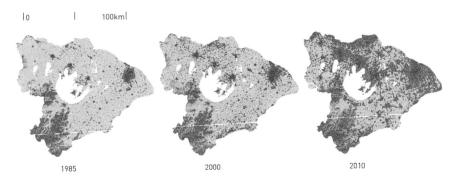

Fig. 2.5 Urbanization in Tai Lake Basin, the main part of the Yangtze River Delta. In-situ urbanization is still ongoing, but city-centered urbanization is becoming way more significant. *Source* Pan et al. (2015). Elaborated by author

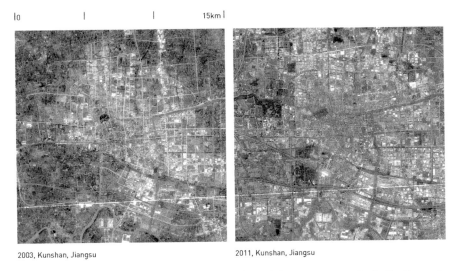

Fig. 2.6 City-centered development in Kunshan: large infrastructure, usually in the form of a regular grid, is imposed on the existing pattern of the agricultural field. *Source* Google earth

and designated special zones in and around key central towns are increasing, and "economies of scale and agglomeration are playing a more important role in urban formation" than the economies of distribution and in-situ (McGee et al. 2007: 143).

The process of involution has been fundamentally reversed in the Yangtze River Delta: more than 80% of the smallholders' income was from the industrial and service sectors in 2007 (Source: China's Agriculture Yearbook, 2007). However, spatially, it shows a sheer privilege of efficiency and unawareness of the agricultural economy and its territory. The construction of infrastructure and urban development usually takes the form of an orthogonal grid, and the affected villages are erased. Urbanization consumes highly productive land and causes severe environmental problems. Moreover, the agricultural and territorial layer is being destroyed.

2.1.6 The Limits of Concentration and the Search for a New Paradigm

If the urbanization stages before the 1990s could be characterized as decentralization in both urban and rural areas, after the 1990s is more a concentration in urban areas with a less-noticed but yet significant decentralization in the countryside (Fig. 2.7). The concentration of urbanization, as a result of the concentration of industry, is approaching a limit—the expected benefit brought by concentration is becoming ambiguous.

The positive effect on economic growth brought about by the concentration of industry is becoming less and less evident. Although the industry in the Yangtze River Delta is among the strongest in China, its productivity has been stagnating or declining since 2005 (Lu 2015; Yang 2014). The advantage of its industrial estates and economic zones is questionable: based on an investigation of 88 China National High-Tech Industrial Development Zones in 2011, the Yangtze River Delta presents

Fig. 2.7 Industrial park in the agricultural field, Tangqi, 2016. *Photo* by author

the slowest growth in China in total industrial output, total industrial added value, net profit, and export, and a negative growth in tax.[29] In 2005, the Yangtze River Delta's industrial concentration had already reached a critical level: profits had been falling as concentration increased (Chen et al. 2008: 84). Research also shows decreasing scale effectiveness, meaning the current scale of the industry is large enough to start to compromise its productivity (Lu 2015: 25). It is a tendency toward an industrial version of involution: the higher the concentration, the lower the profit rate.

The cities are located in the historically most fertile area, thus the concentration of new construction after 2000 within and around existing cities, especially along the Shanghai-Nanjing and Shanghai-Hangzhou urban axes, accelerates the occupation of the most fertile cultivated lands. The Yangtze River Delta is among the most important and modern agricultural zones in China, and its entire territory falls within the zone of optimized development in the National Planning of Sustainable Development of Agriculture (2015–2020).[30] In the Tai Lake Basin, the main part of the delta, the percentage of cultivated land dropped 0.33% per year from 1985 to 2000, and an incredible 1.50% from 2000 to 2010. From 1985 to 2000, the annual net loss of prime cultivated land was 5500 ha. From 2000 to 2010, it reached 20,810 ha, despite the implementation of the Total Dynamic Balance of Farmland in the policy of the Land Management Law, issued by the Chinese government in 1999, and the method of farmland gradation conversion coefficient in 2006. The result is that the arable land per capita was only 0.04 ha, much lower than the security value of 0.053 ha recommended by the Food and Agriculture Organization of the United Nations (Su 2014). Rice production (Shanghai, Jiangsu, and Zhejiang) dropped from 29.29 million tons in 2000 to 26.15 million tons in 2015.[31] The serious loss of farmland has become an obstacle to the sustainable development of these areas (Long 2009; Su 2010).

Concentration was expected to reduce pollution; however, recent studies show the opposite. According to Chen (Chen et al. 2015), an increase of 1% in GDP per unit of land causes 2.08% more industrial wastewater, and 1.04% more industrial soot and dust, which partially supports the "Pollution haven hypothesis"[32] (Chen 2015: 1650). Zhang and Wang's research (Zhang 2014) based on data from 283 cities in China from 2002 to 2011 discovers a double relation: concentration leads to the

---

[29] Source: *The report of the analysis on the statistics and comprehensive development of the national high-tech industrial development zones in 2011*, Torch High Technology Industry Development Center, Ministry of Science and Technology.

[30] The National Planning of Sustainable Development of Agriculture (2015–2020) categorizes three types of zones for agricultural development according to their priority from high to low: zone of optimized development, zone of appropriate development, and zone of protected development.

[31] Source: Source: Statistical Yearbook from 2001 to 2016, National Bureau of Statistics of People's Republic of China.

[32] The pollution haven hypothesis posits that, when large, industrialized nations seek to set up factories or offices abroad, they will often look for the cheapest option in terms of resources and labor that offers the land and material access they require. (Levinson and Taylor 2008). There are also studies by Chinese scholars trying to prove that foreign direct investment causes the "pollution haven" effect. (He 2010).

aggravation of pollution, and pollution depresses further concentration through the increasing environmental costs and falling productivity.[33]

New policies are shifting from concentration to decentralization. In 2014, The National Plan of New Type of Urbanization (2014–2020) was published, which admits the disadvantages of extensive development, and intends to shift the focus of development from large cities to small cities and towns. It relaxes the limits set by *hukou* in towns and small cities while defining and strictly controlling the *hukou* system in large and extra-large cities. While the plan still spatially presents a system of cities and city regions in China and centers on the role of cities as "engines of growth", integrated urban–rural development has been stressed: the construction of universal urban–rural infrastructure, the improvement of rural facilities, etc. The official Regional Planning of The Yangtze River Delta (2009–2020), published in 2010, and The Plan of Development of the City Agglomeration in Yangtze River Delta, published in 2016, share the same principle: The hierarchy of the city (city regions) is defined, and the structure of cities is spatially planned in nodes, belts, and axes. Precise numbers of the incoming population for each city are indicated, the functionality of which is doubtful given earlier experiences. The development of rural areas focuses on the improvement of infrastructure, facilities, and environments. It is ambiguous to limit the growth of (large) cities while emphasizing their role in all aspects of development. In general, the official urban policies intend to optimize the current model on both the urban and rural ends but do not yet provide a real alternative.

At the same time, another action called "constructing beautiful countryside" has emerged from local practices, especially in Zhejiang Province, since the end of the 2000s. In 2013, the construction of "beautiful countryside" was clearly stated in the No. 1 Central Document Focus on Rural Issues. In the same year, Chairman Xi said:

> Even when the urbanization rate reaches 70%, there will still be 400 to 500 million inhabitants living in the countryside. The countryside cannot be one of barrenness, one of the left-behinds, and the old home in the memory. The modernization of agriculture and the construction of new villages must develop, only the development at the same pace can complement each other; (we) have to promote the integration of the urban and the rural.[34]

The most famous and successful three cases of constructing beautiful countryside in Zhejiang Province are Anji, Xianju, and Jiangshan (Huang et al. 2012: 144). All three cases are located in the hilly area with a spectacular, already "beautiful", natural environment. The construction of "beautiful villages"[35] is also accompanied by the "rural construction movement," which often concentrates on specific villages

---

[33] Some research shows the irrelevance between concentration and pollution, especially in the long term (Yan 2011). And others show that the concentration could reduce pollution to a certain extent (Liu and Song 2013; Zhang and Dou 2013). In general, it is fair to say that at least the positive effect of industrial concentration on reducing pollution is not clear.

[34] Source: 中国低碳经济网 (2013). 习近平: 建设美丽乡村不是"涂脂抹粉". [Online] Available at: http://news.sohu.com/20130723/n382304985.shtml [Accessed 18.11. 2022].

[35] The "construction of villages" is strongly influenced by the village construction practice of Japan and Korea from the 1970s, the "Rich and Beautiful Countryside" movement from the 1990s, and the regeneration of villages and land readjustment in Germany and the Netherlands from the 1950s (Chang et al. 2012; Huang et al. 2012; Wei and Shao 2016).

which have a relatively high-quality landscape and where the rural character and traditional aesthetics of China's rurality are strongly present. Accordingly, programs to reactivate the countryside usually include restaurants, home hotels, and nursing homes, to exploit conventional rural beauty. Behind this practice is an ideology of taking rural areas as the complement to the urban, an ideology that intends to eliminate the urban–rural dualism, but ironically with that exact dualism embodied in itself. On the other hand, a new type of "beautifulness" starts to draw attention and is both criticized and appreciated. In many parts of the territory, the public waterfront can be reached by every inhabitant, despite the pervasiveness of the multi-storied smallholders' houses (Fig. 2.8); industrial estates are surrounded by the "garden" of the rural landscape; at the same time, peaceful and large natural spaces can still be found within the densifying urbanity (Fig. 2.9).

The work *New Territory* by avant-garde photographer Ouyang depicts not only the combination of an unusual aesthetic in architecture but also the potentially high-quality environment in the foreground (Figs. 2.10 and 2.11). The series *New Water Town of Jiangnan* manipulates the images with an unrealistically still reflection of the buildings on the water—a ghost of the traditional sense of beauty that could still be perceived around the new construction. A similar angle can be found in Tang's work *Growth* which describes the new images produced in the urbanization process and the relation between the new and pre-existing elements.

The most recent movement is the construction of "characteristic towns". The 13th Five-Year Plan (2016) aims to accelerate the development of small and medium-sized cities and characteristic towns and is detailed as: "By 2020, develop about 1000 characteristic and vibrant towns (which are specialized in) tourism and leisure, trade and logistics, modern manufacture, education and technology, traditional culture, and beautiful (environment) and high livability".[36] It keywords the livable environment, traditional culture, all-around infrastructure, and public amenities, and especially the robust and specialized industry and business, the integration of agriculture and peasants, the complete network of infrastructure, and public amenities in both urban and rural environments. It could be seen as a rediscovering of a "tradition" of China's dispersed urbanization, the first trial with new types of production and innovation. Due to the lack of explicit guidelines, the practice of construction of the characteristic towns tries to learn from western examples, such as Greenwich (US) as a town of finance, Hershey (US) as a town of chocolate, Silicon Valley (US) as a town of high-tech innovation, Vichy (France) as a town of leisure and health, Langethal (Switzerland) as a town of textile, and Grasse (Switzerland) as a town of wine (Zhang 2016). Would it be a better organized and facilitated in-situ urbanization and rural industrialization that avoid the homogeneity, disorder, and waste of the past? Or would it be, as tendencies show, a different version of real estate development that simply repeats a quantitative expansion and densification of towns?

---

[36] Source: The Guiding Advice of Accelerating the Development of Beautiful Characteristic Small Cities and Towns, National Development and Reform Commission, 2016; The Notice Of The Development Of Characteristic Small Towns, Ministry of Housing and Urban–Rural Development, National Development and Reform Commission, Ministry of Finance, 2016.

Fig. 2.8 Multi-story peasants' houses at waterfront, Tangqi, 2016. *Photo* by author

Fig. 2.9 Baitayang Lake, "hidden" in the dense urbanity, Tangqi, 2016. *Photo* by author

Fig. 2.10 Jangnan Yang, *New territory*. Ouyang (2016), published by Jiazazhi press

Fig. 2.11 New water town in Jiangnan, *New territory*. Ouyang (2016), published by Jiazazhi press

2.2 The "Third" Space

The third space is a concept used in different disciplines. Its theory emerges from the socio-cultural tradition in psychology identified by Lev Vygotsky. In social geography, Soja, inspired by the work of Lefebvre, introduces the concept of the "Thirdspace" in which "everything comes together... subjectivity and objectivity, the abstract and the concrete, the real and the imagined, the knowable and the unimaginable, the repetitive and the differential, structure, and agency, mind and body, consciousness and the unconscious, the disciplined and the transdisciplinary, everyday life and unending history" (Soja 1996: 56–57). It is a space that encompasses the Firstspace, which is the physical aspect of space, and the Secondspace, which is its ideological and imaginative aspect. In the theory of media arts, the third space idea has been used by media artist Randall Packer as a networked and shared space distinct from the hierarchical structure and top-down media (the first space) which connects the entirety of the local and the remote (the second space).[37]

The territory of the Yangtze River Delta can be read as three physical spaces: the core cities, the remote and solitary villages, and a vast, continuous, and dispersed urbanity in between–the third space. The third space is a result of the geography of the territory, the historical agricultural economy, the process of in-situ urbanization, and rural industrialization, a space of *desakota*.[38] It covers the majority of the Delta, but it is often neglected in current research and policy—a space that is vast, complex in its physicality, neglected or overlooked mentally, and created and experienced by saturated types of human activity.[39] It is crucial to rediscover and understand the third space, especially when searching for a "new type of urbanization" (新型城镇化) and in view of the tendency to adopt a more decentralized model -. The following section provides a brief introduction to the third space in the Yangtze River Delta, and the current research and policies covering it.

---

[37] See the website of The Third Space Network by Packer: https://thirdspacenetwork.com.

[38] The term *desakota* was introduced by McGee to describe "regions of an intense mixture of agricultural and nonagricultural activities that often stretch along corridors between large city cores" (McGee 1991: 7). He explains the composition of the term: "the use of a coined Indonesian term taken from the two words kota (town) and desa (village) was adopted after discussions with Indonesian social scientists because of my belief that there was a need to look for terms and concepts in the languages of developing countries that reflect the empirical reality of their societies. Reliance solely on the language and the concepts of Western social science, which have dominated the analyses of non-Western societies, can lead to a form of "knowledge imperialism." In this text, I have used the term *desakota*, which can be used interchangeably with *kotadesa* (McGee 1991: 23–24).

[39] In this sense, the understanding of this physically "third" space precisely requires the use of the Thirdspace theory. McGee in his research on China (McGee et al. 2007) constantly quotes and refers to Lefebvre's concept of "production of space" (Lefebvre 1991).

2.2.1 The Introduction of a "Third" Space in the Yangtze River Delta

The Yangtze River Delta can be perceived as a continuous urbanity covering a large territory within a 250 by 250 km frame.[40] In 93% of cases, the distance between built-up areas is never more than 15 km or 15–20 min by foot.[41] This continuity exists since the 1930s when Fei discovered that all villages in the Delta were never more than 20 min away by foot (Fei 1939a: 28). This is comparable to other city-territories in the world, such as Flanders, Belgium, and a part of the Veneto region in Italy. However, the population density is generally 1552/km^2,[42] not only much higher than the one of the European city regions such as Flanders (557/km^2) and the Veneto region (260/km^2) (Fig. 2.12), but also much higher than many definitions of "urban", such as that of the urban core (1500/km^2, 1000/km^2 in Canada and US, by OECD[43]), urban areas in Japan (500/km^2), and in China (1500/km^2).[44] Furthermore, 15 million more inhabitants are expected to settle in the Delta from 2014 to 2020.

About 7% of the built area is disconnected urbanities–mostly remote villages located in the periphery of the delta and hilly areas. Within 93% of the continuously built area, the urban cores are mainly Shanghai, Hangzhou, Suzhou, Zhangjiagang, Wuxi, Changzhou, and Nantong, most of which are located in the northern part of the Delta and along the Shanghai-Changzhou axis. In between the two spaces, a carpet of urbanized areas with denser nodes covers more than 60% of the total built area, a third space (Fig. 2.13).

This third space can also be recognized by the distribution of density. The highest population densities are located along the Shanghai-Changzhou, Shanghai-Zhangjiagang, and Shanghai-Hangzhou axes, and the settlements in between the axes have a lower but "urban" density (200 to 2000/km^2). The third space has good infrastructure. The entire Delta is connected by a grid of mobility arteries consisting

---

[40] There are multiple popular ways defining the Delta: the first is based on the geographic physicality, especially the figure of the Tai Lake Plain, including Shanghai, the southern part of Jiangsu, and The northern part of Zhejiang; the second is based on the definition of the Shanghai Economic Zone approved by the central government in 1992, which extended the former boundary to part of northern Jiangsu and part of central Zhejiang; the third is based on the urban agglomeration defined by the official planning of urban agglomeration of the Yangtze River Delta by the Ministry of Housing and Urban–Rural Development, which also includes the eastern part of Anhui Province. Taking into consideration the multiple definitions, this study chooses a 250 × 250 km frame.

[41] Source: GIS data, 2010, National Earth System Science Data Sharing Infrastructure. Elaborated by Author.

[42] Here the density is of the Tai Lake watershed in 2010, which is 36,895 km^2 and the majority of the 250 km-by-250 km area (Su et al. 2016). The density of Shanghai, Jiangsu, and Zhejiang, covering 213,081 km^2 of land and much larger than the 250 km-by-250 km frame, is 748/km^2. (Source: Statistical Yearbook 2016, National Bureau of Statistics of People's Republic of China, elaborated by author).

[43] Source: Definition of Functional Urban Areas (FUA) for the OECD metropolitan database September 2013.

[44] Source: Inventory of official national-level statistical definitions for rural/urban areas, International Labour Organization, 2015.

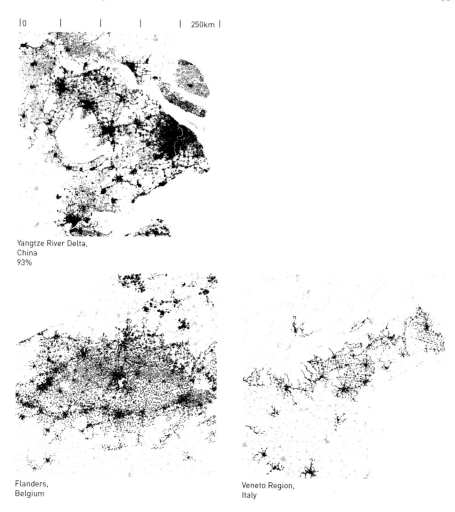

|0 | | | | 250km |

Yangtze River Delta,
China
93%

Flanders,
Belgium

Veneto Region,
Italy

Fig. 2.12 Comparison: built area in Yangtze River Delta, Flanders, and Northern Italy. Black: the continuous urbanity (less than 1.5 km from each other, or 15 min on foot). Gray: the discontinuous urbanity. *Source* GIS data, 2010, National earth system science data sharing infrastructure China, and CORINE land cover 2000, European environment agency. Elaborated by Author

of railways, highways, and national roads, with the grid becoming denser as it nears Shanghai. It is also connected with an immense capillary system of local roads, including county level, town level, and village level roads, which support the substantial number of dispersed towns and villages. In the third space, especially, county roads often function as "boulevards" in rural areas—a structural network able to carry public transport, cargo, and private vehicles. The Yangtze River Delta also has a rich water system. Countless rivers, ponds, lakes, and canals form an isotropic network that used to function as the main mobility artery. It is a staggering network to the road

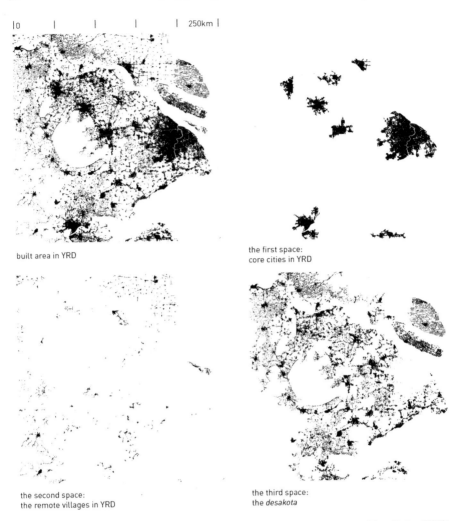

Fig. 2.13 First, the second, and the third space: a spatial analysis of the Yangtze River Delta (YRD). *Source* GIS data, 2010, National earth system science data sharing infrastructure China. Elaborated by Author

network, instead of following the latter. The dispersed urbanity in the third space has had a very close relation to the water since its inception: smallholders' dwellings are often built along the water, the villages around bridges, and the towns at the junction of infrastructural waterways. The third space usually has a myriad of local facilities such as medical services and schools due to its high population density (often more than 1000/km^2 in the villages). The local mobility and water network and the diffuse local facilities function as a universal support base for this type of space.

The third space shows wide variety. It has the full spectrum of forms of water: in the areas adjacent to Tai Lake, especially in its east and south, the water tends to take the form of lakes and ponds, enclosing the land into islands. The areas close to the seaside display long parallel ditches perpendicular to the coastline, while other areas have different canal networks and branches of waterways dividing and penetrating the fields. Paddy fields are the main type of agriculture, with orchards, nursery fields, fishponds, and other types of agriculture found in other areas. The morphology of the villages and towns also differs according to the form of water and type of agriculture. Nevertheless, the similarity of all the elements lends a strong universality to the third space.

The third space has a unique landscape, which has the character neither of the cities nor the traditional villages: it is densely built with 2–5 story houses but without tower blocks. The inhabitants' income is drawn mostly from urban activities and an important agricultural sector. Tractors, buses, trucks, cars, motorbikes, and bikes use the same road. Vast paddy fields and water surfaces are found in places, garden-like vegetable and rice fields in others. There is ambiguity about the character of the third space. The main question is: Is this an urban or traditional rural space? This ambiguous feeling is also found in *Zwischenstadt*, which "consists of 'fields' of various uses, construction forms, and topographies" with "both urban and rural char-acteristics", in which "the ratio of open landscape and built-up areas have frequently been reversed; the landscape has changed from being an all-inclusive 'background' to being a contained 'figure'" (Sieverts 2003: 2–3).

Such a landscape is not planned centrally but as a consequence of diffuse authority. Since the main part of the third space is defined as "rural", it is often neglected by urban planners. It is, rather, the result of an enormous diffuse operation that is decided and organized locally. Before 1949, collective agricultural work, along with the first rural industries (starting from the 1930s), was organized by villages or subvillage groups. During the Maoist era, not only the collective agricultural and industrial production but all aspects of people's lives were organized by communes (on the scale of today's towns) and brigades (on the scale of today's villages). After 1978, the towns and villages inherited autonomy from the communes and brigades and started to manage their collective economy on collective land. Today, the local autonomies are set in a global economic system design. The seven administrative levels–nation, province, city, county, town, village, and household—are each contracted by the higher level, but compete within the same level.[45] This decentralized power—each county, town, and village is simply a company of its own—results in an incredibly diffuse development. This can be seen clearly in the distribution of industry: the national—and provincial-level economic zones are absent in most of the third space, because of the city-centered policy. County- and town-owned industrial parks, small ports, village industries, and all levels of industry of industries are scattered through the third space, collaborating and competing. This diffuse industrialization brings diffuse densification to the existing dispersed urbanities, which is still an ongoing process.

---

[45] Cheung explains this economic system as the true reason for the miracle of China's economy. A detailed explanation on county level is provided in his paper *The Economic System of China* (Cheung 2008).

The third space is not the main objective of urbanization today, and in most of these spaces, the percentage of migrants is low. In light of the expected 15 million newcomers by 2020, the third space could potentially be the answer for the next stage of urbanization, a new paradigm. Its productive agriculture dispersed but also dense population, well-developed infrastructure, facilities, and mixed landscape of urban and rural provide a unique opportunity to construct a new type of city. Therefore, it is critical to understand and describe the third space.

2.2.2 State of Art: Research and Policies

The Yangtze River Delta as an exemplary case of urbanization has been intensively studied for the form, transformation, and mechanism of its urbanization (Che et al. 2011, 2015; Ge 2015; Gu 1991, 1992, 2001; McGee 1991; Sun et al. 2011; Yao 1992, 1998; Yu 2005, 2017; Zhou 1991). Since the 1990s, this body of research has discovered and tried to describe a continuously urbanized area that is far beyond the boundary of the cities in the Yangtze River Delta.

It starts with the notion of "corridor" (or axis) which is inspired by Gottmann's Megalopolis concept[46,47] and has become a mainstream tradition. It extracts a linear form of urbanization from a broader urban territory, leaving the rest behind. The MIR Nanjing–Shanghai–Hangzhou in the Yangtze River Delta has been presented in two ways: one is "by drawing a circle whose radius extends fifty kilometers from an incorporated city along the given transportation corridor" (Zhou 1991: 99); the other is by "using the administrative boundaries of core cities located in the fifty-kilometer-wide zone on each side of the given corridor" (Zhou 1991: 99). The intercity and high-speed railways, highways, and the coast are the corridors or axes of urban development (Gu 2001: 332). Further studies on how to define the axes and their strength have been conducted, using detailed data analysis and advanced mathematic models (e.g., Sun et al. 2011).

Another notion also extensively used is the classification of the cities, or the urban system (城镇体系). In the MIR concept, the cities are simply classified into 5 levels according to their population. This 5-level "pyramid" system (Gu 2001: 334) has become a standard reading of the urban space. It is interesting that, in the 1990s, the interaction between the different levels of the cities was not emphasized. Yao stated that the 2nd level cities (Hangzhou and Nanjing, the biggest cities in the Delta after Shanghai) have an unobvious influence on their surrounding counties and cities. In contrast, third-level cities are rapidly developing, depending on an "endogenous"

---

[46] Zhou notes "a number of large cities along transportation corridors" as the common feature of the *kotadesai* zone and the megalopolis. He also notes that the main difference between the two is that the former is "located in the richly endowed agricultural areas…the limited amount of farmland has been a strong repelling force for agricultural population." (Zhou 1991: 89).

[47] Gottmann listed the "urban constellation around Shanghai" as one of the six global megalopolises. (Gottmann 1976).

power.[48] From 1995 to 2000, and after in the twenty-first century, the model of cities as singular poles started to be replaced by the core/axes and the city agglomeration models, due to the large-scale/global investment and construction of infrastructure, especially the Z-shaped mobility axis of Nanjing–Shanghai–Hangzhou-Ningbo. The urban space in the Delta has been divided into Megalopolitan areas and the hinterland between. (Gu 2001: 334). In 2003, the blue book, "The Report of Urban Development in China (2002–2003)" was published, which officially introduced the idea of urban agglomeration (组团式城市群) in China.

The official policy and planning are coherent with those two notions. The Plan of the Development of Yangtze River Delta (2009–2015–2020), published in 2010, introduced spatial planning of "one core and nine belts" (Fig. 2.14). The core is the city of Shanghai, and the nine belts are located along the main mobility infrastructure and the water features (Yangtze River, Tai Lake, the coast, Hangzhou Bay, and the Great Canal). The cities were categorized into six levels. The system and distribution of different industries were designed according to this spatial structure and hierarchy of cities. The most recent plan, The Plan of Development of the City Agglomeration in Yangtze River Delta (2016–2020–2030), published in 2016, defines a spatial structure of "one core, five agglomerations, and four belts" (Fig. 2.15). The core is the city of Shanghai. The five city agglomerations are located along the urban axis Nanjing–Shanghai–Hangzhou–Ningbo, the most significant belt in the previous planning concept. The four belts are along the main infrastructure, the coast, and the Yangtze River. In most of the plans, the definition of the urban system is related more to economic and demographic aspects, which inevitably leads to a rough and diagrammatic representation of the spatial form of the agglomeration.

Another body of research recognizes the extensive area with a mixed rural and urban character besides the urban centralities—the third space. The most significant is the concept of *desakota* by Terry McGee, "which are regions of an intense mixture of agricultural and nonagricultural activities that often stretch along corridors between large city cores." (McGee 1991: 7). It recognizes several types of spaces in the territory: the major cities, the peri-urban, *desakota*, densely populated rural, and the sparsely populated frontier. While in the peri-urban area people commute to the city core daily, *desakota* implies a more independent and in-situ development. It has also been recognized by John Friedmann, who has been studying Asia and China for decades and considers the in-situ urbanization in the rural area as "truly transformative". This thesis has followed and also challenges comparable concepts by both Western and Chinese scholars, including "metropolitan interlocking regions" by Zhou (1991) and "peri-urbanization" by Webster (2002), among many others. Similarly, Yao used the term "urban continuity" to describe the phenomenon of the shrinking hinterlands between cities and the formation of a continuous industrial urban belt in the Suzhou-Wuxi-Changzhou area. (Yao 1998: 2). Gu directly

---

[48] Although Yao listed Suzhou, Kunshan and other cities in this category, he explains endogenous power over external investment and other conditions as the primary reason for the fast development. It echoes Friedmann's perspective on China's urbanization that, "instead of globalization…urbanization as an evolutionary process that is driven from within, as a form of endogenous development." (Friedmann 2005: XVI).

Fig. 2.14 General planning (one core and nine belts), The plan of the development of Yangtze River Delta, 2010

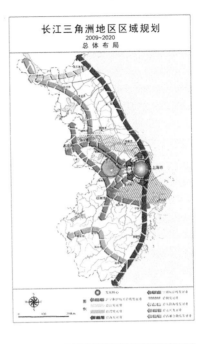

Fig. 2.15 Spatial frame of the urban agglomeration (one core, five agglomerations, and four belts), The plan of development of the city agglomeration in Yangtze River Delta (2016)

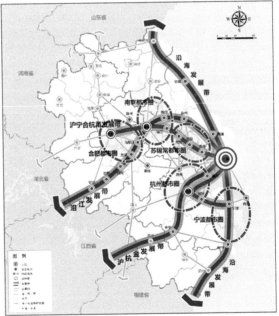

implemented the term "megalopolis" (Gu 2001)—literally in Chinese as the Yangtze River Delta continuous metropolitan area—to describe the delta as one entity with a distribution of city groups and their hinterlands.

Contemporary technology, data collection and analysis, and new measures enable researchers to significantly extend the scope to a territorial scale, to look at the vast third space in a more accurate and detailed way. Quantitative methods are being employed: the scale-rank law to measure the tendency toward concentration; ROXY law to measure the degree of population density; the influence formula and mathematic model built to measure the attraction between cities (Sun 2011). The result shows that the middle- and small-scale cities are growing faster than the large cities, which means the urban system or hierarchy tends to be more horizontal and diffuse. It shows a tendency since 1986 toward a more evenly distributed population among the cities, and a large group of "secondary cities" after Shanghai, in which no clear leader could be identified. It also shows the rising influence of the secondary cities, some of which were the hinterland of Shanghai, but which today are becoming the main influencers on the local scale and reclaiming their hinterlands. Che's research based on satellite images (Landsat MSS/TM/ETM from 1980 to 2007) confirms this diffusion process (Che et al. 2011). It illustrates three stages of spatial pattern in the Yangtze River Delta: the first stage from the 1980s to 1995, when three urban cores (Shanghai, Nanjing, Suzhou-Wuxi-Changzhou) were the evident hot spots of development; from 1995 to 2000 when a great number of hot spots started to appear and merge with the previous cores; and from 2000 when the different spots started to merge rapidly into belts/agglomerations. What draws attention is the analysis using the lacunarity index, which shows that the high-lacunarity (low in hinterlands between urbanities) area extracted itself from 2000 to 2007 from the limit of the Z axis (Nanjing–Shanghai–Hangzhou–Ningbo) and extended into the northern and southern part of the Delta, displaying the diffuse tendency of the urbanization (Che et al. 2011: 451). Based on a marginal K function data analysis, another similar study defines the degree of concentration tendency and defines the urban agglomeration via the economic space (Ge 2015). The peak of this tendency represents an optimal situation–a balance between the advantage brought by the concentration of economic activity, infrastructure, and facilities, and the disadvantage brought by overcrowding, traffic jams, and degradation of the environment. The area in the Yangtze River Delta that has the highest level of concentration/attraction is a vast continuous area in the middle of the Delta between the large cities. For example, Shanghai and Nanjing, the two biggest cities, and their immediate surroundings in the region are not part of this area. The figure of this area shows a completely new condition different from the core-belt model. The most recent research by Yu and Zhou (2017) defines three types of development models: leapfrogging, edge-expansion, and infilling.[49] The result is that, different from the Pearl River Delta, the Yangtze River Delta "had a decreased

---

[49] Leapfrogging refers to the model in which new patches of developed land are not adjacent to existing patches; edge-expansion represents the pattern in which new patches of developed land grow out of the fringe of existing patches; infilling refers to the pattern in which new patches of developed land are surrounded by existing patches. (Yu and Zhou 2017: 2 of 18).

proportion for edge-expansion and an increased value for infilling" (Yu and Zhou 2017: 7 of 18), which represents an opposite tendency to the center-periphery model, with a regeneration of the center plus diffuse and isotropic urbanization in the territory (Fig. 2.16). The diffuse urbanization presented by this variety of indexes and measurements also leads to (or is confirmed by) diffuse increases in rurality. The research by Zhang (2014) shows a consistent decrease in the rurality index in the Yangtze River Delta, a concentration of agriculture-oriented villages on the northern edge, a small area of agricultural land in the southern part of the Delta, and a vast area of mixed villages and industrial and service-oriented enterprises in the middle.

Attempts have also been made to investigate in detail the local transformation of the third space. The outcome uncovers variety in the settlements and spaces of the rural area in the Yangtze River Delta, rather than homogenizing all the space into the urban or the rural (Wu 2015). Moreover, the outcome presents a more gas-like and "pervasive" urbanization (Zhuo 2011) in the Yangtze River Delta. Based on Google Earth satellite images, Statistical Year Books, and other data, Wu claims a transition of the rural space in the Yangtze River Delta (especially in the coastal villages in Zhejiang) "from low-density and diffuse, toward high-density and homogenized" represents "a unique tendency towards non-agriculturalization, dispersion, and suburbanization," and "it makes a clear difference from most of the studies based

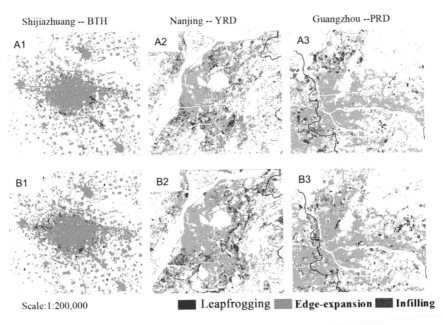

Fig. 2.16 Examples of the different morphological models of urban expansion in cities of the three urban megaregions. A1–A3 are results in 2000–2005, and B1–B3 are results in 2005–2010. (Yu and Zhou 2017: 7 of 18)

on the Southern Jiangsu area" (Wu 2015: 667). She claims a "universal urban situation" of transformation from the rural/village to the suburban condition. (Wu 2015: 668). Two elements were analyzed in detail: the housing area and the roads. In 2010, the average housing area per capita in the rural area of the Yangtze River Delta was 51.6 m^2 (1.5 times the national average in rural areas), while that in the urban area was only 32.1 m^2. In Zhejiang Province in 2013, the length of roads per capita was 41.6 m in rural areas, 3.2 times that in urban areas; the surface of the roads per capita was 33.3 m^2, 1.3 times that in the urban area. Although the discovery leads the authors to conventionally advocate for the polarization of the urban and rural environments as the solution (villages look like villages, contrasting with fast urban development) (Wu 2014), it shows an incredible capacity and "embodied energy" in the rural area. At the same time, it also highlights new challenges in the rural area, for instance, the upgrading and maintenance of such a pervasive network of rural roads, which reveals an "urban" condition. This type of research unfortunately does not yet comprise a significant part of the research on urbanization in the Yangtze River Delta.

In a conclusion, describing the continuous urbanity in the Yangtze River Delta has been a constant effort for decades. Today, new materials including data and satellite images, and new technology and methods for measuring and analyzing enable scholars to collect and process information in a much larger territory—from the central cities to the third space—and to draw more detailed and precise conclusions. Although the classic description, especially reflected in the current urban policies in China, still intends to read the urbanization from a hierarchic and core-axes model that may still be predominant in the Yangtze River Delta, the new outcomes reveal a more territorial, diffuse, isotropic, and horizontal condition on different scales of urbanization. This provides new evidence of and relevance to the school of McGee, Friedmann, Zhou, Webster, etc. However, very few studies try to reveal the physical and detailed nature of this condition. What is the materiality of the third space? How was it built? What kind of transformation is ongoing? This thesis intends to add to the research efforts and describes the third space in the Yangtze River Delta from a spatial point of view.

2.3 Starting from the Elements

It's still not usual in the research tradition on China's urbanization to start from elements. However, elementarism is a way of tackling the complexity which characterizes the third space of the Yangtze River Delta. The following section thus presents a brief reflection on the notion of elements from the work of a group of architects and tries to clarify the relevance of this notion to today's territory of the Yangtze River Delta. This thesis looks at two aspects: the repetition of the elements and the potential of elements as design objects. It proposes using elementarism as an analytic and design tool to decipher the complexity and to propose a new spatial agenda for the third space. Through this different notion of elements, the author also proposes a methodology with specific hierarchy, scales, and terminology.

2.3.1 The Notion of the Element(s)

It is not new to try to understand and describe the contemporary city based on its elements, as the contemporary city is written, different from 19th-century cities, "note by note: a city of fragments, endless parts that come to rest next to one another in random ways" (Viganò 1999). The notion of elements is strongly presented in the *Progetto di Suolo* by Secchi, *La città elementare* by Viganò, *The Form of the Territory* by Gregotti, *Learning from Las Vegas* by Venturi, *The Living City* and *The Broadacre City* by Wright, The *No-stop City* by Branzi, *The Generic City* by Koolhaas, and many others.

It is not just a necessity due to the lack of a legible or known composition in contemporary cities, but also a "step-back" action "that tries not to close within known fixed categories whatever eventually will have to emerge, though it may not necessarily do so" (Viganò 1999). By retreating to the basics and into detail, studying the elements of the fragmented reality generates a minimum but fundamental certainty before interpretation, and creates the maximum space to construct new possibilities from it. This retreatment is especially pertinent in today's China, where urbanity is a mixture of contemporary operations and a long history of habitation, where the various ideologies, policies, movements, and practices overlap with each other constantly, and where in the decades of the rapid urbanization process the new proposals are implemented before the previous one delivers results.

2.3.2 Two Aspects of the Elements

First, the elements are repetitive. They are clear forms that are repeated "yet within this city that we interpret as fragmented and heterogeneous, we also see that certain objects and particular materials repeat: car parks, shopping centers, detached housing developments, sports grounds, etc.: materials that can be named and are finite in their forms, have precise, defined characteristics, and are organized into various types of sequences" (Viganò 1999). And this generality "can be examined as a whole from different viewpoints, … bringing to light relationships between them that are not initially evident". The repetition of certain elements can also generate a certain degree of fragmentation, such as the consistency Venturi found in the repetition of streetlights in Las Vegas with the uneven rhythm of other objects in the background (Venturi 1972: 20).

The recognition of the generality and consistency of elements is critical to the third space of the Yangtze River Delta. While most of the current studies and urban policies address the specificity of each village and attempt to bond the strategy of urbanization with it, the territory is constituted by a rather limited but common collection of materials and forms. Taking a train from one side of the Yangtze River Delta to the other, you would see a continuous space with dispersed housing, factories, infrastructure, farmland, vegetable fields, water ponds, billboards, etc., and be impressed by not just the amount of repetition of the elements, but also the extreme simplicity and the genericity of each element: the countryside roads are simple concrete surfaces about 5 m wide, bordered by tall fir trees; the small dikes dividing the paddy fields; peasants' houses, almost identical in their volumetric configurations, orientation, type of roof, material, and windows; electric towers across the fields; identical basic vegetable winter gardens with their plastic film cover.

The repetition of some of the elements, especially the modern ones, brings genericity. Some parts of the territory can be read as a rural version of *The Generic City*, as the repetition of elements oriented to functionality, efficiency, autonomy, and competitiveness dilutes the already un-evident relations between them. Koolhaas reports the contemporary city through elements in different scales, organized in a continuous text without any paragraphing: airports, hotels, atriums, communities, new town, waterfront, …each of these elements appear in an independent piece of text. The genericity is rendered by the lack of dynamics between things: "The Generic City is… a place of weak and distended sensations, few and far between emotions, discreet and mysterious like a large space lit by a bed lamp" (Koolhaas 1994, page 1250). The public space, the street, the green space, and the infrastructure—the classic structural objects—have lost their roles in holding the city together. Branzi agrees, "the ever-weaker connections that unify the city, architecture and the world of objects" (Branzi 2006, page 10). On the other hand, the identity and history of each element are lost through duplication: "as it is more abused, it becomes less significant." Each element is transforming in the same direction: sedated, smooth, lack of urgency, neutral…The Generic city describes the result of this "systematic application of the unprincipled" as "varied boredom" (Koolhaas 1994, page 1262). An increasing majority of the elements in the third space of the Yangtze River Delta are lacking conventional historical or aesthetic meaning. With genericity as a theme being overlooked, the new operations such as the "characteristic towns", which intend to present specific characters for each case, often become a pursuit or invention of non-existent specificity and lead to generic construction in the end.

While the meaning of some of the elements is *diluted* by repetition, in the third space of the Yangtze River Delta, the meaning of the others is stagnating or even being intensified. The vast scale of the territory, which provides relatively large spaces between elements, also gives a certain autonomy and resilience to the pre-existing elements. There is no doubt that the lack of design in the construction of the elements, and the lack of "articulation of the spaces" (Secchi 1986) in the rural area contribute greatly to the rural–urban divide. There is also no doubt that the territory is receiving rapidly new elements with new logic and relations, and the coherence is interrupted episodically by undecipherable conditions in the field, such as scattered high-density housing developments, highway nodes, and industrial estates with colossal buildings. There is no doubt that the phenomenal societal transformation is changing the meaning of the elements, as Gregotti witnesses, "the fundamental aspects of the morphological impulse": large-scale human migration, a profound transformation in third-world countries, and recent technological developments, especially the last one "have led to an even more acute and dramatic deepening of the chasm concerning the pre-existing anthropo-geographical structure that often constitutes the invaluable testimony of a culture" (Gregotti 1981). But because of such a social transformation, the density of elements and complexity of the territory has reached an interesting level. The vast territory provides relatively large spaces between objects, maintains the legibility of the existing, and provides room for the new. For instance, in some parts of the Delta, the pre-existing trees inside the villages, the orchards which have been planted broadly since the 1980s, and the new trees brought in by the new infrastructure and new construction in the villages orchestrate a strong presence and continuity of green space, which was not the main characteristic of the Delta. The newly built peasants' houses are slicing some of the vast paddy fields into pieces, which creates a feeling of gardens in between the houses while productivity is maintained. The migrants settling in the villages have stimulated new housing typologies. Different "solutions" are invented in different parts of the Delta but can still be interpreted as an evolution of the old village houses. If the intensity and variety of the elements in Fei's era were closer to the one of the *No-stop City* by Archizoom Associates (Archizoom Associates 1971) the fullness of today's territory of the Delta reminds one more of the drawing of the *Patchwork Metropolis* by Neutelings et al. (1989), as the ancient organization of certain elements is still clearly present and functioning.[50]

Second, the elements are genetic. The landscape is presented by the elements. As a result, it can be read through the elements, and eventually transformed through the transformation of the elements. According to Gregotti, the landscape as an ensemble of elements is also a "signifying environment" that has "a historical origin because a decision was made that lead to its selection or rejection". In this sense, every landscape is specific: even if they are geographically similar, "our perception of landscape is always determined by history, and we ceaselessly recreate geography through our cultural experience as users". The elements function as signifiers due to their repetitive presence in daily life, their determination of the "plots" of the landscape as the grid for the division of the soil, and the particularly dense meaning some of them have. The elements function physically: "the overall anthropogeographic environment has a materiality". The meaning of culture and nature refers to the physical environment, to its formal structure. History is the "fundamental support of its formal structure". The materiality of the elements is significant: the territory conceptually becomes a "field", an operational unity, an object for design. The planner can detect a number of ensembles (of elements), read the structure of the field by understanding the intersection between the ensembles, and define the signifying contents displayed in the ensembles. He/she can thus operate on those signifying contents to intervene in

---

[50] The intensity and variety of elements are not only supporting but also are attributed to a certain level of population density; the density of the majority of the Yangtze River Delta, even in many rural villages, has reached $1000/km^2$, higher than the $900 km^2$ of Randstad in the beginning of 1990s, when the "patchwork metropolis" concept was born.

the territory and eventually the society. It is a process of understanding the territory to design the territory through elements; this is a process that began with the first appearance of humans within that territory.

This notion of the element as not only a cognitive but also an operational object is significant for the third space of the Yangtze River Delta, where urbanization, especially the top-down part of it, is charging the territory with new elements while neglecting the potential to transform the pre-existing elements. Here, the new elements, seemingly independent of the territory, often ignore the pre-existing elements to that they are subordinated. The territory as the ensemble of elements itself is perceived in this case as a blank canvas instead of a "palimpsest", according to Corboz, which is "so heavily charged with traces and with past readings" (Corboz 1983). For example, the erasure of villages due to the construction of new infrastructure leads to "new village" construction, which is supposed to be compact and concentrated but in reality "unconsciously" becomes a diffused densification of the existing village system. The trees and arable land of the erased villages are abandoned, leading to a decline of the ecological environment, which also accommodates the new arrivals. The flower and grass beds in the neighborhood (Fig. 2.17) and along the new roads (Fig. 2.18) are presented as modern qualitative elements. They are sometimes transformed into vegetable beds by the inhabitants—these vegetable beds are not just physical elements but should be read also against their background, with the rising consumption of vegetables per capita in urban areas, and the declining consumption in rural areas.

Fig. 2.17 Flower and grass beds in the neighborhood is used as vegetable fields in the rural area of Hangzhou, China. *Photos* by author

Fig. 2.18 Flower and grass beds along the road is used as vegetable fields in the rural area of Hangzhou, China. *Photos* by author

Understanding the territory through certain "forgotten" elements, which could be pre-existing but also structural and genetic, is thus urgent; the integration of those elements into the design process is therefore also urgent. As Secchi claims, the territorial structure, or the "frames", can be imagined necessarily by the "singular character of everyday interventions", and we must "imagine interventions that will complete the frame, slowly modifying it, bending it until it takes on a new meaning until a new inhabitable space has been built" (Secchi 1991).

2.3.3   A Methodology of the Elements

An individual element, except "figures" like an airport or a volcano atoll, usually does not function in a territorial role alone; a methodology is needed to process the collection of elements. Many terms such as ensemble, sequence, typology, layer, structure, system field, etc., have been used to describe the organization of multiple elements. A clear methodology must be chosen for the study of the Yangtze River Delta to avoid the confusion of various intentions behind each term.

In *Morphology, Material* (Gregotti 1985:2–3), Gregotti constructed a hierarchy with three levels: the morphogenetic elements/materials, the architectural form, and the morphology of the territory. The morphogenetic elements or the materials are "any aspect of reality capable of (consolidating) producing architectural forms", thus they can be physical or not: natural forms, building regulations, land subdivisions, stylistic memories, the traditional typological models of geometry, the modes of production—a material is "any aspect of reality capable of (consolidating) producing architectural forms". And those architectural forms are "presented as models to be imitated, interpreted or concretized, or as elements whose hierarchy and composition can be reorganized through new formal organizations and new presences in a specific field". In this sense, both the elements and the forms they produce are morphogenetic. More than just rules of aggregation or the shape of the territory, the morphology is "a capacity to generate and preside the design of complex systems", through "morphogenetic elements".

Here, the architectural form could be compared to the complex urban material as Viganò defines it. The morphogenetic feature of the architectural form enables it to produce the territory exactly as a morphogenetic element, although it is itself composed. Furthermore, the emphasis on the complexity of certain elements, subtracting the inner mechanism of their subcomponents and naming them as forms and typologies is not the main intention of this study. This study has a relatively narrow scope to describe the vast third space of the Yangtze River Delta by studying a selection of its very simple elements, which are rather pale in their complexity. The intangible elements, which could certainly deliver the physical form of the territory, are excluded also here, due to the "retreating from interpretation" approach. Therefore, the first level of the study of elements could be called elements—physical elements, including both the simple and the complex.

Above the level of individual elements (and the architectural form), Viganò introduces the notion of the layer in a structural sense. The layer of a single element not only displays its repetition but makes it possible to reorganize complex relationships within an ordered series of simple relationships. Only from the layer, the order of the elements that are decomposed from the fragmented objects of the territory can be revealed, i.e., the order that recomposes the element into an entity. From there, the underlying or potential territorial structural role of certain elements can be discovered. The description of the layer can be done simply via mapping—an action that does not presuppose the existence of a structure, an action very different from the current studies and policies that intend to immediately interpret and impose structures on urbanization. The layer also creates a territorial reading that overcomes the rural/urban division, which is unusual in today's studies.

The third level is the system. According to Viganò, the concept of a system "can refer to concepts of identity and belonging that is, of the nameability and recognizability of each element and its belonging to a group, a whole, a family" (Viganò 2000). This group, for example, an ecological system, consists of various elements and their layers. Certain rules and a structure integrate them into the whole. Since the rules and structure can be designed by an architect, so can systems. For example, the school "system" conceived by the Soviet disurbanists was composed of layers of auditoriums, libraries, small schools/classrooms, accommodation in the forests, etc., which are the elements decomposed from the traditional complex of schools. Its functionality depends on and interacts with other systems, such as the universal road and railway system, the forest system, and the housing system. The current category of systems, such as the economic, the ecological, the hydraulic, the mobility system, etc., has few types and is often related to specific expertise. It cannot include or provoke, as in the example of the school, new relations and qualities utilizing the existing systems. Consequently, this study does not intend to evaluate and intervene in the conventional systems, as intensive efforts in this field have already been undertaken by other researchers. Rather, it attempts to imagine new systems and be inspired by systems invented by "utopians" that were not implemented due to certain historical and contextual constraints, or that were indeed implemented in underlying but overlooked ways. This thesis examines these issues in the specific context of the third space of the Yangtze River Delta.

In conclusion, the study of elements consists of three levels: the elements, the layer, and the system, which follow Viganò's terminology. The elements are discovered via intensive fieldwork and case study on the scale of 3.2 km by 3.2 km (or 2 miles by 2 miles, as in the Broadacre City model of Wright). The elements include very simple ones such as the concrete surface, more complex ones such as the peasant houses, "typologies" such as the water management unit similar to the one discovered by Fei, and the combination of orchards and fishponds. The elements are mapped in layers on a scale of 3.2 km by 3.2 km and 50 km by 50 km, which constitutes a part of the Yangtze River Delta. The mapping of elements on such a scale is unusual in the state of the art, and it is expected to produce a new and particular type of knowledge. Finally, new systems are proposed on the scale of 50 km by 50 km in abstract manners but depicted back in detail on a scale of 3.2 km by 3.2 km, to clarify the physical transformation of the tangible elements, and the quality this brings to life.

It has to be clear that some of the elements discovered on the local scale can be found or applied to the 50 km-by-50 km territory, such as the industries. Others cannot, such as the *yu* (圩) water management unit or the orchards. Mapping all the industrial elements on a scale of 50 km by 50 km conveniently confirms some of the discoveries on the local scale. However, by mapping all the water elements on the scale of 50 km by 50 km, a diffuse water network with similar intensity in every part of the territory could be illustrated, as the elements of *yu* function in a significant portion of the territory. One could easily imagine a variation of *yu*, carry on a similar exercise of research and design based upon it, and repeat this for the local orchards and territorial green layer, and many others. The shift in scale does not compromise the legitimacy of the local elements on the territorial scale. On the contrary, it highlights their territorial role by integrating them into a diverse and rich unity. This mechanism is presented more clearly in the systems, in which the common quality of the "different" elements within a territorial layer is used, as the accommodation of the students in a forest does not demand the particularity of the trees. This mechanism can also be imagined extending to the entire Yangtze River Delta, a frame of ± 250 km by 250 km, which requires an extensive study on a much broader scale. This study intends to propose this possibility for further discussion and tries to offer the first step in this direction.

References

Archizoom Associates (1971) No-stop city, residential parkings, climatic universal sistem, *Domus* 496, 1971 mars. [Online] Available at: https://tribologie.wordpress.com/2014/10/23/archizoom-no-stop-city-domus-496-mars-1971/. Accessed 18 Nov 2022

Branzi A (2006) Weak and diffuse modernity: the world of projects at the beginning of the 21st century. Skira, Milano

Chang J, Zhu D, Feng S (2012) 德国村庄更新及其对我国新农村建设的借鉴意义[J]. 建筑学报 2012(11):71–73

Che Q, Duan X, Guo Y, Wang L, Cao Y (2011) Urban spatial expansion process, pattern and mechanism in Yangtze river delta. Acta Geogr Sin 66(4):446–456

Cheung SNS (2008) The economic system of China. 2008 Chicago conference on China's market transformation

Fei X (1939a) Peasant life in China. E. P. Dutton Company, New York

Fei X (1939b) Peasant life in China: a field study of country life in the Yangtze Valley. George Routledge and Sons, London

Fei X (1948) 鄉土中國. Guancha, Shanghai

Friedmann J (1981) The active community: toward a political-territorial framework for rural development in Asia. Econ Dev Cult Change 29(2):235–261

Friedmann J (2005) China's urban transition. University of Minnesota Press, Minneapolis

Friedmann J, Douglass M (1978) Agropolitan development: towards a new strategy for regional planning in Asia. In: Lo F-C, Salih K (eds) Growth pole strategy and regional development policy. Proceedings of the seminar on industrialization strategies and growth pole approach to regional. Pergamon Press

Gottmann J (1976) Megalopolitan systems around the world. Ekistics 243(February):109–113

Gu C (1991) A preliminary study on the division of urban economic regions in China. Acta Geogr Sin 46(2):129–141

Gu C, Zhang M (2001) Study on the characteristics and dynamics of Yangtze delta megalopolis. Adv Earth Sci 16(3):332–338

Guldin GE (ed) (1997) Farewell to peasant China: rural urbanization and social change in the late twentieth century. M. E. Sharpe, Armonk, NY

He W (2010) FDI的 "污染天堂假说" 检验: 基于中国东部和中部的证据 [J]. 当代财经 2010(6):99–105

Huang P (1990) The peasant family and rural development in the Yangzi delta, 1350–1988. Stanford University Press, Stanford, California

Huang S, Wu Q, Pan C (2012) 国外乡村发展经验与浙江省 美丽乡村 建设探析 (Development experience of foreign rural area and analysis on the beautiful countryside building in Zhejiang Province). Huangzhong Architecture 华中建筑 2013(5):144–149

Lefebvre H (1991) The production of space. (Nicholson-Smith D, trans.). Blackwell, Cambridge, MA

Levinson A, Taylor MS (2008) Unmasking the pollution haven effect. Int Econ Rev 49:223–254. https://doi.org/10.1111/j.1468-2354.2008.00478.x

Li W (1996) 人多地少国家发展高效农业的原则. 中国农村经济 5:14

Liu X, Song D (2013) 城市产业集聚对城市环境的影响. 城市科学 212(3):9–15

Mao Z (1998) 毛泽东读社会主义政治经济学批注和谈话 (上册). 中华人民共和国国史学会. Translated by author. Cited in Zhang, H. (2018) 毛泽东构建新型工农城乡关系的探索与启示. 毛泽东思想 2018(02). [Online] Available at: http://rdbk1.ynlib.cn:6251/Qk/Paper/649913#anchorList. Accessed 18 Nov 2022

McGee TG (1991) The emergence of *desakota* regions in Asia: expanding a hypothesis. In: Ginsburg N, Koppel B, McGee TG (eds) The extended metropolis: settlement transition in Asia. University of Hawaii Press, Honolulu

McGee TG, Lin G, Marton A, Wang M, Wu J (2007) China's urban space: development under market socialism. Routledge, New York

Meisner M (1977) Mao's China and after: a history of the people's republic. The Free Press, New York

Neutelings WJ, Sulsters W, van Wesernael P, Winkler E (1989) De transformatie van de Haagse zuidrand, Stedebouwkundige studie verricht in opdracht van de Dienst Stadsontwikkeling en Grondzaken van de gemeente s G ravenhage, in het kader van het stadsplan—1988–89. In: Geurtsen R, Verschuren P, Kwakkenbos G (eds) Stadsontwerp in s-Gravenhage: stedebouwkundig ontwerpen¡. Faculteit der Bouwkunde, Delft

Ouyang S (2016). New territory. Jiazazhi Press

Pan P, Yang G, Su W, Wang X (2015) Impact of land use change on cultivated land productivity in Taihu Lake Basin[J]. Sci Geogr Sinica 35(8):990–998

Sit V (2015) 新中国成立以来城市化与城市发展的回顾. [online] HPRC 中华人民共和国国史网. Available at: http://www.hprc.org.cn/gsyj/yjjg/zggsyjxh_1/gsnhlw_1/jjgslw/201110/t20111019_162379.html. Accessed 22 July 2017

Soja E (1996) Thirdspace. Blackwell, Malden (Mass.)

Spate OHK, Learmonth ATA (1967) India and Pakistan. A general and regional geography. Methuen, London

Su W, Chen W, Guo W, Ru J (2016) The occupation of river network by urban-rural land expansion in Taihu Basin, China. J Nat Resour 31(8):1289–1301

Sun G, Wang C, Xiao L, Jin F (2011) Temporal-spatial characteristics of evolution of the urban system in the Yangtze river delta. Resour Environ Yangtze Basin 20(6):641–649

Wei Y, Shao C (2016) Research about the experiences and strategies of beautiful country in typical regions of the world. Sustain Dev 可持续发展 6(3):189–195

Wen T (2009) 三农问题与制度变迁. China Economic Publishing House, Beijing

Wittfogel KA (1957) Oriental despotism. A comparative study of total power. Yale University Press, New Haven

Wu K (2014) The transition of rural space in Zhejiang Province. Dev Reform Res 288(5):1–24

Yan F (2011) 产业集聚发展与环境污染关系的考察. 科学学研究 2011(1)

Yu W, Zhou W (2017) The spatiotemporal pattern of urban expansion in China: a comparison study of three urban megaregions. Remote Sens 9(1):45. https://doi.org/10.3390/rs9010045

Zhang K, Dou J (2013) 集聚对环境污染的作用机制研究. 中国人口科学 2013(5):105–116

Zhou Y (1991) The metropolitan interlocking region in China: a preliminary hypothesis. In: Ginsburg N, Koppel B, McGee TG (eds) The extended metropolis: settlement transition in Asia. University of Hawaii Press, Honolulu, pp 89–111

Zhou Y, Huang X, Xu G et al (2016) The coupling and driving forces between urban land expansion and population growth in Yangtze river delta. Geogr Res 35(2):313–324

Chapter 3
Elements: A Case Study of Tangqi

Abstract The first part of this chapter elaborates on the reason we chose the town of Tangqi, located in the southwestern along the Yangtze River Delta as the subject for our case study. My intensive fieldwork in Tangqi uncovers several elements which are further explained in the second part. A micro-story of these elements and their history, social-economical background, functionality, and a description of their physical performance accompanied by photos and interviews complete our analysis. Thus, the rationality, the transformation, and the challenges related to each element are laid bare. The third part deals with the urban–rural split by the elements and reflects on the articulation of the elements, the notion of public space in a rural and urban setting, and a comparison between the two.

Keywords Case study · Description of the elements · Micro-story

3.1 Introduction

The following paragraphs explain a selection of spatial elements in Tangqi. Those elements are discovered in a 3.2 by 3.2 km (or 2 by 2 miles) area (Figs. 3.1 and 3.2) that is the identical area covered by the Broadacre city model by Fran L. Wright. The selected area is on the outskirts of Tangqi, a mixed space with a bit of everything: water, agriculture, different scales of industries, and historical and modern settlements. It is meant to be a biopsy sample of the ordinary tissue of the third space of the Yangtze River Delta.

The explanation of the elements is not only through the spatial analysis including mappings and descriptions of their current physical performance, but also through micro-stories[1] of the elements with their history, social-economical background,

---

[1] A similar concept to micro-story, micro-history, was originally developed in Italy in the 1970s. According to Giovanni Levi, one of the pioneers of the approach, it began as a reaction to a perceived crisis in existing historiographical approaches (Levi 1988). Levi worked with Carlo Ginzburg (another creator of micro-story) and Simona Cerutti on Micro-story, a series of micro-historical works. The concept of micro-story is also used by designers. Bernardo Secchi and Paola Viganò, in the construction of the structural plan of Antwerp, address the necessity of constructing the micro-story of the elements:

© The Author(s), under exclusive license to Springer Nature Switzerland AG 2023   51
Q. Zhang, *The Elemental Metropolis*, The Urban Book Series,
https://doi.org/10.1007/978-3-031-36409-9_3

|0 | | | | 250km|

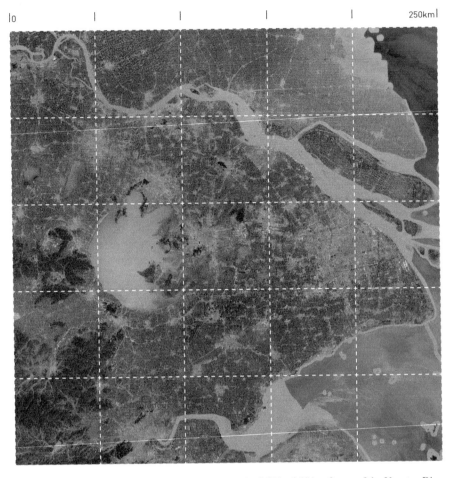

Fig. 3.1 Case study of Tangqi: the 3.2 by 3.2 km in the 250 by 250 km frame of the Yangtze River Delta

and functionality. The work is a result of intensive fieldwork and the collection of first-hand and grounded information.

In the impossibility of constructing systematic knowledge of the past, depicting the roots of certain images, and tracing the genealogy of the issues which substantiate them today, the micro-story starts from an inevitably piecemeal kind of knowledge highlighting the elements that become useful for a more general discussion (Secchi and Vigano 2009: 25).

|0 | | 3.2km/2miles|

Fig. 3.2 Case study of Tangqi: the aerial image of the 3.2 by 3.2 km area in 2016. *Source* Google earth

3.2 Case Study: Tangqi

Tangqi is an ordinary town that rarely catches any attention in urban planning. While Kunshan, one of the first big-scale industrial platforms in China and one of the most unique and successful areas in China's urbanization, Tangqi is a town of the average in the Yangtze River Delta. Its administrative area is 79 km², close to the town average (105.5 km²) of Zhejiang Province; it has a population of 79,990 and a population density of about 1012/km², close to the average of the third space (823/km²). The migration percentage is 16.4%, close to the Zhejiang Province average of (13.2%), and much lower than areas like Kunshan (63.6%).[2] While industrial output

---

[2] The population density is based on the Sixth National Population Census of the People's Republic of China, 2010, and elaborated by the author.

far outweighs agricultural production, agriculture still contributes 5.9% to the GDP of Tangqi (2013), close to the average of Zhejiang Province (4.2% in 2015), and far higher than large industrial platforms such as Kunshan (0.8% in 2008). At the same time, it is not yet a suburb of big cities. It is critically distant to the city center of Hangzhou (one of the major cities of the delta) is 20 km, and 13 km to Linping which is the closest subcenter of Hangzhou. In some sense, Tangqi is autonomous as the great number of "ordinary towns" in the Yangtze River Delta. While industrial cities like Kunshan receive all kinds of advantages from policies at all levels, while satellite towns like Linping are governed by the urban planning of their affiliated big cities and consequently receive programs and resources, while protected villages like Anji are promoted as a test area for the construction of "characteristic towns" and ruled by specific planning and design, Tangqi, as an ordinary town of non-exceptional quality, must find its path like most towns in the third space.

Tangqi is located in the Tai Lake watershed. It has the typical geography of the Delta: an enormous number of lakes, rivers, and canals. The resulting geographic and natural environment and the type of agriculture show many common characteristics of the territory. The transformation of the rural area around this Tangqi has a history of more than 600 years and is considered by some scholars to be a complete example of rural economic and social evolution in the twentieth century (Dong 2014, p. 2). The town has a rich culture, which generated a relatively broad collection of literature and documents helpful to this research. Moreover, the author's father was a member of a commune in Tangqi during the Cultural Revolution and worked as a consultant for industries in Tangqi. His rich knowledge and social relations greatly helped in collecting first-hand information from fieldwork and interviews.

3.2.1   Element 01: Yu 圩

Rice has been the most important crop in the Yangtze River Delta for a long time. The paddy field demands precision irrigation: the water level must be guaranteed and checked regularly under the "eye" of the plant—the upper joint of the leaf with the stalk. At the same time, it demands intensive fieldwork: each laborer can manage only about 0.4 ha or 6 mu paddy field. Therefore, the water must be managed collectively.

A particular water management unit was explained in detail in Fei's description of Kaixiangong Village in the 1930s, and even today, this unit still exists and functions in many parts of the Delta, including in Tangqi (Figs. 3.3 and 3.4). This unit is a plot of land surrounded by water, with a large circular dike on its perimeter to protect the land against flooding. The plot is called *yu* 圩. The size of each *yu* varies greatly according to the natural form of the water. For the center of the *yu* to be irrigated, the whole *yu* resembles a dish—high at the periphery and low in the center. A series of small dikes are built parallel to the border of the *yu*, to construct a cascade system allowing the water to reach the center gradually without causing flooding. The dikes slice the entire *yu* into small plots, and the land is level within each plot. Intermediate dikes are used to divide the *yu* into parts called *jin* 瑾 when the *yu* is too large. The

|0 | | | 3.2km/2miles|

Fig. 3.3 Water surface in the 3.2 by 3.2 km case study area, 2014. *Source* Google earth, elaborated by author. "a" and "b" two *yu* (圩) next to each other. See details in Fig. 3.4. Dark gray: water surface between the different *yu*. Light gray: fishponds inside of the *yu*

land value decreases dramatically from the perimeter to the center of the *yu*, as they are the last plots to be irrigated and the first to suffer flooding.

Two systems of ditches are made to manage the water: one for irrigation (section A) and the other for drainage (section B) (Fig. 3.5). Along the perimeter, there are several pumping points to pump the water from the lower river into the *yu*. The water is pumped into ditches that are perpendicular to the perimeter and sent to the middle of the *yu* following the artificial topography of the *yu*. Every small plot is connected to the ditch by an inlet. The most peripheral and therefore the highest plots are irrigated first. Their inlets are opened, and the water is prevented from flowing down. When these plots are irrigated, their inlets are closed, and the water is allowed to flow to the next level of the cascade. The lowest plots are irrigated last. Another

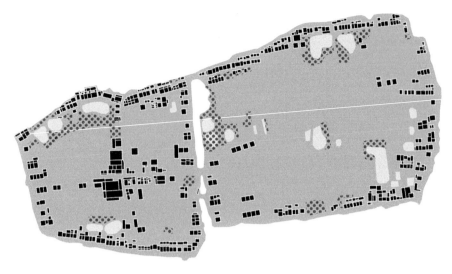

Fig. 3.4 Diagram of two *yu* (圩) next to each other. Dark grey: land, light grey: fishponds, red: orchards, black: buildings. *Source* Google earth, elaborated by author

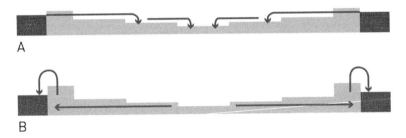

Fig. 3.5 Diagram of the two systems of ditches in sections

system of ditches connects the lowest point of the *yu* (or *jin*, if the *yu* is too large) to its perimeter, where several pumping points are built to send the water to the river over the dike when there is flooding. These two systems are still in operation today.

The large and intermediate dikes function as the road in the *yu* but also leave space for other uses. The buildings are built on the large dike at the perimeter of the *yu*. In Tangqi, the houses can be found on the perimeter of a *yu*, while in other parts of the territory, the houses are more concentrated on one part of the dike, as shown by Fei. The rest of the dikes are used for vegetable fields, mulberry trees for the silk business, and in the case of Tangqi, the loquat orchards. From the 1980s, when the peasants had more freedom to use the land fishponds became a popular business. They are built around the *yu*, often as part of the dike, leaving the center of the *yu* for agriculture (Fig. 3.6).

|0 | | | 3.2km/2miles|

water
fishpond
agriculture
orchard
housing

Fig. 3.6 Overlapping of the water with other elements through mapping: fishponds, agricultural land, orchards, and housing, 2014. *Source* Google earth. Elaborated by author

As a water apparatus, the *yu* creates significant water-related value. The various water surfaces in and around the *yu* are crucial for the stability and development of the ecosystem. This value includes aquatic products, water management (supply, regulation, storage, and flood control), purification (reductions in nitrogen and phosphorus levels), temperature regulation, carbon fixation, oxygen production, and biodiversity maintenance. Among them, water management accounts for the largest share due to its capacity to create a balance between the normal and maximum allowed water levels of the river, paddy field, and fishponds, and its infiltration capacity. Moreover, the ecological value of the *yu* area, the indirect part of which is far larger than the direct part, is very high, because of the structural role of the water ecosystem in the entire natural ecosystem of the region.[3]

The dike at the southern perimeter of the *yu* (Fig. 3.7, 1A)

The buildings are on the dike. Many of these dikes are still soft, but their transformation into hard (stone, concrete) constructions is taking place gradually. The inhabitants can reach the water, to wash their clothes or their food in traditional ways, via small, often perpendicular stairs that lead into the dike. Thus, they informally enclose part of the dike and create a semi-domestic space in front of each house. This space is used for storage, parking, vegetable plots, drying food and clothes, washing, temporary constructions, toilet, etc. Private handrails, fences, and trees are used to mark its border. The wastewater is often directly discharged into the main water body.

The dike at the eastern/western perimeter of the *yu* (Fig. 3.7, 1B)

The buildings are perpendicular to the dike making it more difficult to create a semi-domestic space. Sometimes, decorated handrails in the style of those found in ancient Chinese gardens are added, delineating the waterfront as a public space. This level of detail is found often in urban areas. Stairs leading to the water are for public use.

A ditch for irrigation (Fig. 3.7, 2A)

The water level in the ditch is equal to or higher than the water level in the plot or paddy. The ditch is concreted, which makes the dike also a small road. The ditch is used regularly (and informally) for sewage, although a public sewage system is often constructed when the road is concreted over. The pollutants and garbage that land in the ditch via the rainwater, along with the domestic and industrial wastewater, cause severe land pollution.

---

[3] See the study by Yan et al. (2015), in which the aquatic ecosystem service value is assessed. The study attempts to define a precise economic value for each of the ecoservices of the *yu*, which is questionable in terms of the numbers, but still offers a general picture. While the ecological value of a *yu* differs according to its location in the Delta, the ecological value of the *yu* in Tangqi is among the highest in the *yu* areas.

Fig. 3.7   Photographic inventory of the water elements. Photos by author

A drainage ditch (Fig. 3.7, 2B)

The water level is visibly lower than the water level in the plots. Rapeseed, loquat trees, and vegetables are planted along the concreted dike. Maintenance is kept up in some places, as shown in this photo taken in an "exemplary agricultural zone" but is poor in many other places, partially due to the lack of interests of the villagers in agriculture.

Vegetables planted on the dike by a metal fabricating enterprise (Fig. 3.7, 3)

Besides the small number of vegetables grown wherever possible by the peasants, there are also relatively large and well-managed vegetable fields developed by rural industries to feed their workers. These fields are located around the factory and often on the slope of the dike.

The small plots (Fig. 3.7, 4)

The plots are used as paddy fields and sometimes used for growing water vegetables such as Zizania Latifolia.

Loquat orchards on the dike (Fig. 3.7, 5)

The loquats planted on the dike are for domestic consumption, and this fruit has been traditionally extremely popular in Tangqi. The loquat trees are found on the dikes of every *yu* and give a strong consistency to the landscape.

Dikes within the village (Fig. 3.7, 6)

The villages are often built near the bridge connecting two *yu*. Peasants' dwellings are built on the dike, on both sides of the narrow river that runs through the village. In contrast to the fruit trees that are grown for economic purposes, beautiful old trees indicate the "center" of the village and provide a cozy and public space where people can meet.

The dikes, as the fundamental part of the water management apparatus, are a space. The large dikes at the perimeter of each *yu* vary in width from 3 to 15 m and are often the largest public space in the village. The occupation of the space is mixed and dynamic and can be a road, a productive space, parking for inhabitants, and visitors, or a meeting point. It is a public space but open for private use. And although it is a private space, it can never be completely enclosed. It is a collective space—a space almost impossible to imagine in the contemporary urban environment, where spaces are often strictly divided into public and (gated) private.

3.2.2 Element 02: Fishponds and Orchards

The agriculture in Tangqi is famous for its combination of sericulture, fruit, and fishery (Fig. 3.8). The shrinking share of grain in agricultural production has been a common trend in the Delta since the Ming Dynasty when the market economy started to emerge. Tangqi has high precipitation, hot summers, and wet and cold winters, conditions are perfect for mulberry and loquat orchards, and the geography of the plain provides abundant water surfaces for fisheries.

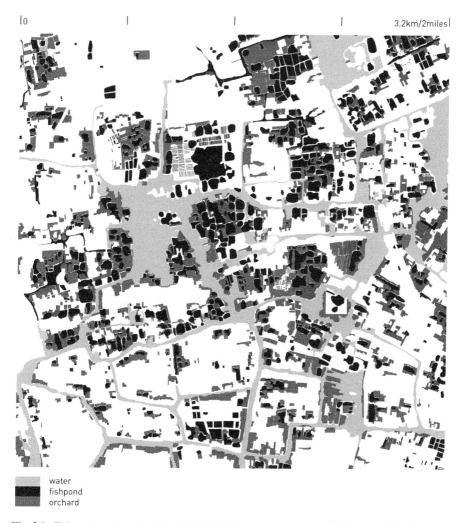

water
fishpond
orchard

Fig. 3.8 Fishponds and orchards in the 3.2 by 3.2 km case study area, 2014. *Source* Google earth, elaborated by author

After the modern sericulture industry was introduced in China at the end of the nineteenth century, mulberry trees occupied sometimes more than 70% of the total arable land in some parts of the town in the 1930s, while the paddy fields were reduced to about 10%. After the 1930s, because of market changes, and especially after the war, the mulberry orchards shrunk dramatically. During the Maoist era, the surface area of mulberry orchards remained at 100 ha. After the reform, and despite increasing industrial pollution, sericulture recovered as a result of booming consumption brought by urbanization. In 2005, mulberry orchards occupied 391 ha in Tangqi.

The loquat is another specialty of Tangqi (Fig. 3.9). The inhabitants started to grow loquat during the Sui Dynasty (581–618), and this fruit became popular during the Tang Dynasty (618–907). After the war and during the Maoist era, the loquat orchards were reduced from 800 ha in the 1950s to 400 ha in the 1970s due to pollution and government policy. After the 1980s, these orchards started to recover and occupy 1000 ha in Tangqi today, including large production bases. The fishponds followed a similar trajectory. Since the 1950s, increasing industrial pollution, triggered by the low topography, has been the main reason for the reduction in fish production. Another reason is government policy that privileged grain production during the Maoist era and led to the transformation of some fishponds into paddy fields. Urbanization stimulated the production of fish after the 1980s. Today, there are 420 *mu* (or 28 ha) of fishponds.

The fishponds and orchards are often located on the dikes, which results in a special ecology and landscape. During the middle of the Ming Dynasty, the linear arable plot was invented and established in the Yangtze River Delta. The low-lying part of the land was dug up to make fishponds, and the soil was recycled to build

Fig. 3.9 Drone photo of the fishponds and orchards in the 3.2 by 3.2 km case study area. Photo by author

the dikes. Substance and energy are exchanged between the two systems with the fishpond and the orchards on the dikes. Combining the land and water ecosystem significantly prolongs the food chain and enhances biodiversity. The manure of the fish, the decomposed water plants, and fallen leaves, in the form of sludge, are used as a natural fertilizer for the orchards. The trees protect the pond from wind and cold and consume the sludge of the pond, preventing the latter from becoming silted. The most famous type of orchard is mulberry, which expands this circle: the mulberry is used to feed the silkworms which produce silk, and the manure of the silkworms can be used to feed the fish. This ecological cycle can be expanded by including livestock such as pigs and sheep, and the orchards can include fruit trees, flowerbeds, grass, sugar cane, etc. (Fig. 3.11).[4] The implementation of this system has been constantly adapted in Tangqi: at the beginning, it was the version with mulberry orchards and fishponds, and due to the decline of the silk industry, today, the mulberry has been replaced by loquat trees.

In this part of Tangqi, almost every family grows loquat. The loquat trees are usually 3–5 m tall, with long broad leaves, which gives the open and flat territory a very closed atmosphere. Loquat trees are evergreens, with beautiful white or yellow flowers in autumn or early winter. The trees start to be productive after 3 years and fruit for decades after that. They appear in different typologies: linear ones on the border of the *yu*, patches, or as vast spaces with masses of trees near large lakes. Overall, the loquat orchards can be perceived as a continuous landscape stretching over the territory with both esthetic and economic value. However, this landscape is at risk. Each family living there works both in industry and in agriculture, maintaining the orchards. Due to the limited time and labor that can be spent on the orchards, their production is enough only for domestic consumption. The relocation of the peasants to concentrated settlements/cities will lead to the disappearance of the orchards. At the same time, water pollution is causing a rapid decrease in fish production (Fig. 3.10).

Loquat forest (Fig. 3.12, 1)

The orchards can be massive, like a little forest, sometimes large enough to have roads running through. This type of space is unusual in the open and flat Yangtze River Delta.

Planted dike (Fig. 3.12, 2)

The orchard can also run linearly along the dike. It creates a dense but low facade along the water. They could appear as lines on the two sides of the dike, or be planted as a lane leading up to the entrance of the house.

---

[4] The system is widely studied. See a different version of this system in the Pearl and Yangtze River Delta (Zhong and Cai 1987), and an incredibly extended historical version of the mulberry-dike-fishpond in Huzhaou, Yangtze River Delta (Zhou 2013). In 2017, the Huzhou Mulberry-dyke & Fishpond System was proposed as a site of Globally Important Agricultural Heritage Systems (GIAHS), FAO.

Fig. 3.10 Mr. Wei explains the decline of the production of fish, which could not be a means of income anymore. The Loquat trees that each family keeps are just several tens. Photo by author

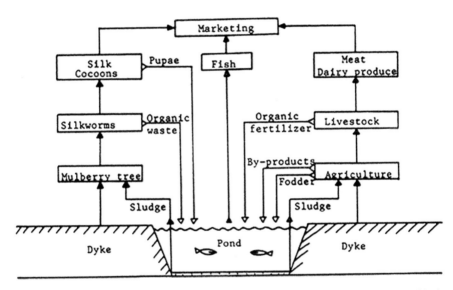

Fig. 3.11 Typical Chinese integrated fish/agriculture/livestock farming system (FAO 1983; for specific examples, see Edwards 1982)

Street trees (Fig. 3.12, 3)

Loquat trees are used as street trees. Visually, the street trees are the spin-off of the "forest" stepping into the artificial space.

Fig. 3.12 Photographic inventory of the elements of the fishponds and orchards. Photos by author

Decorative trees (Fig. 3.12, 4)

Loquat trees are also used to decorate the public space in the village, such as the banks of a water pond. They are mixed with other types of flowering trees.

Fishpond in the "forest" (Fig. 3.12, 5)

Within the forest, paddy fields and fishponds can be found. Some of the most high-quality sites have been developed for recreational fishing.

The fishpond and orchard in front of the forecourt of a row of houses (Fig. 3.12, 6)

The houses have direct access to and a view of the water, sometimes on both sides, which is a luxury and unusual condition when living in the city, but available everywhere here. The loquats are planted all around the pond, which provides privacy for the ground floor of the houses without blocking the views from the upper floors. The inhabitants use the water of the pond for washing or keeping ducks.

Open water surface (Fig. 3.12, 7)

The fishery is practiced in the open water as well. The vast open water surface (*Beiyayang*) creates an incredibly beautiful landscape, hidden in the middle of the densely populated villages by a series of smaller fishponds and orchards.

The hardening of the dikes (Fig. 3.12, 8)

The dikes of the fishponds are usually soft, but today more and more hard dikes made of stones have begun to appear.

The combination of orchards (loquat in this case) and fishponds creates a unique productive landscape. As if by prior agreement, the loquat trees are used for almost all kinds of purposes and in all kinds of spaces: in the fields, along the dike, the streets and the paths, next to the houses, within the enclosed private gardens, etc., which provides a universal language of the village vegetation. This language is used in the public, collective, and private spaces, which blurs the distinction between them—or, to create a feeling of universal collective space, a universality created simply by common economic reasons. The fishponds, although threatened by pollution, create high-quality living conditions for the inhabitants, and there is no cost for maintenance. On the other hand, truly public space is lacking due to the only-for-profit mentality.

It is a diverse landscape. Although this part of Tangqi loquat predominates, another part of Tangqi will have predominantly green plum trees. It is also a changing landscape. During the last century, the proportion of area taken up by orchards, fisheries, and rice shifted dramatically, which shows the incredible flexibility of the territory. Today, with the ongoing urbanization process, an increasing need for agricultural products may stimulate more or different productive landscapes.

3.2.3   Element 03: Roads

Roads in China are classified into national, provincial, county, town, and village roads
according to their managing authority and are classified into four levels according
to their technical specifications. The county, town, and village roads are categorized
as rural roads, managed and maintained by local authorities. In general, the width of
the rural roads is strictly controlled to minimize the consumption of arable land: the
standard county and town roads are more than 6 m wide (excluding the shoulders),
and the village roads are wider than 3.5 m. The surface of the county and town roads
must be either made from asphalt or concrete, while it is recommended that village
roads are made from asphalt or concrete (Fig. 3.13).

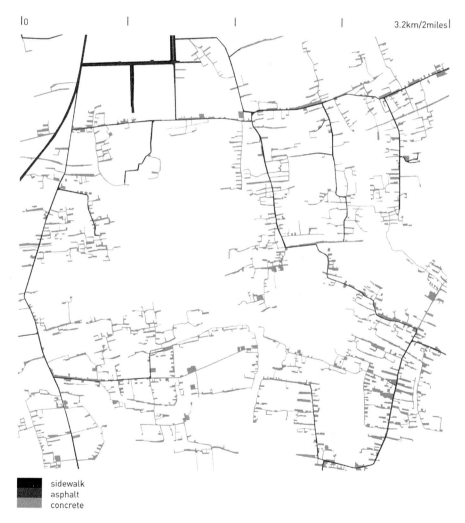

|0 | | | 3.2km/2miles|

sidewalk
asphalt
concrete

Fig. 3.13 Different types of roads in the 3.2 by 3.2 km case study area, 2014. *Source* Google earth,
elaborated by author

In the 1930s, the roads connecting the villages and towns were used mainly for towing boats against the headwind or countering the currents in this area. In 1945, there were only a few hard-surface roads connecting the cities, the rests were loose-surface tracks and trails. During the commune era and after 1956, a national construction plan promoted the construction and upgrade of roads for tractors (called tractor-plowing roads), with a sand-and-stone surface varying from 3 to 4 m to about 10 m wide. Some of functioned as county town roads, others as village roads. Fir trees (in some cases, pine trees) were planted along them. The planting of slim and tall fir trees was very popular in the Maoist era and gave a strong identity to the rural area. In Zhejiang Province in 2003, only 14.5% of village roads were hard-surfaced, and a province-wide operation of concretization was launched as a result of the abundant production of concrete in the province. The fir trees were either cut down or limited to very small spaces along the road due to the maximization of the surface for traffic. In 2009, village roads with hard surfaces connected all the villages in Tangqi. Concrete is the most common material for covering village roads because of its low cost, simple building technique that allows peasants' participation, low maintenance, and long life span. The base of the road is made of fragments of stones and bricks, which are abundant in today's countryside. However, concrete surface cracks are common, while concrete degeneration creates dust. Sidewalks are rare. The forecourts of the peasant's houses are also constructed in concrete, making it a continuous surface with the road. When a road passes a village or a small town, this concrete space is often enlarged for parking, car washing, shops, and informal open-air markets. The rainwater flows into the irrigation ditches, along with pollution and rubbish. New vegetation, such as small pine and fruit trees, is planted interrupting the former setup. The monotonous use of concrete and the general lack of street vegetation create a pale and dusty atmosphere in the villages and lack intention and collectivity.

Urban road types have been introduced to rural areas by industrialization, which renders a strong urban character contrasting with the immediate rural surroundings. In between the plots of industrial estates, roads have been built identical to the ones in the cities: orthogonal grids, asphalt surfaces, gutters, more than four-lane roads, separate bike lanes or an extra space beside the car lanes, and sidewalks in concrete bricks. Camphor trees, planted universally in the cities but unusual in rural areas, are planted on the sidewalks. The new esthetic is not fully respected: peasants occasionally use the tree holes or the flowerbeds along the roads as vegetable fields.

The main road in the industrial estate (Fig. 3.14, 1A)

The road is part of the urban road system. It has a green belt in the middle and sidewalks with camphor trees on both sides. The surface is asphalt. The width is ±20–25 m, with 2–3 lanes for cars in each direction.

Fig. 3.14 Photographic inventory of the elements of the roads. Photos by author

The secondary road in the industrial estate (Fig. 3.14, 1B)

The road is connected perpendicularly to the main road. It has sidewalks with camphor trees on both sides. The surface is asphalt. The width is ±15 m, with 1–2 lanes for cars in each direction, and 1 lane on both sides for parking.

County roads/town roads (Fig. 3.14, 2)

Usually, there are no sidewalks but tall and slim fir trees on both sides. The fir trees are from the Maoist era and give a very strong legibility to the roads. The surface was stabilized with stone and sand during the Maoist era, which was changed to concrete or asphalt afterward. The width for traffic must be more than 6 m, in this case about 8–10 m. The traffic includes buses, private cars, trucks, motorbikes, bikes, and pedestrians, without indication of lanes. No gutters are present; thus, the rainwater flows into the ditches of the fields along the road.

Village roads (Fig. 3.14, 3)

The village roads are simply a plain concrete surface of ±4 m wide. Due to the lack of space, there are few trees along the roads. As the village roads are mostly found on the larger dikes in the field, rainwater flows to the open ditch in the field. A few transformations of the road—adding asphalt cover, replacing the open ditches with underground pipes, planting decorative trees—can be found here and there.

The asphalt surfaces (Fig. 3.14, 1A, 1B, 2)

The country roads or town roads together create an asphalt surface that crosses the middle of the *yu*, due to the lack of space at the periphery of each *yu* (Fig. 3.15). These roads connect 2 *yu* with a bridge that is usually higher than the dike to allow boats to pass. The middle part of the yu is connected in this way and becomes a preferred site for new housing, factories, and public facilities (schools). In contrast, the space at the periphery of each *yu*, the first choice for housing throughout history, has become quiet and marginalized—if the original villages were connected by the grid of waterways, the asphalt roads provide an entirely new mobility system. The new construction sometimes gives the road a continuous facade of activities and an urban atmosphere. It is used as a space for a mix of activities in those parts (Figs. 3.16, 3.17, and 3.18). When school is finished, the street is crowded with buses and waiting parents, and all traffic comes to a halt. Shops and informal markets along the street can attract a large crowd, and sometimes, the entire space is taken up by a temporary street market. Loading and unloading trucks are a frequent occurrence on these roads. It is a shared space for all types of mobility, from pedestrians to trucks.

   The roads could be categorized according to their materials, and each material has its character and organization.

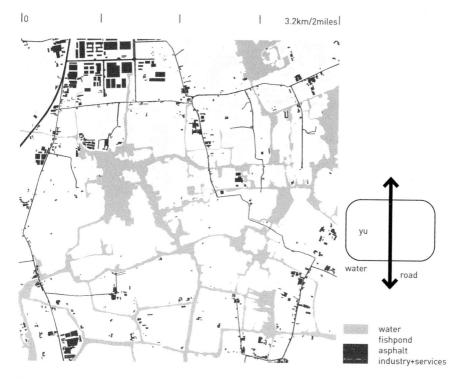

Fig. 3.15 Overlapping of the asphalt surfaces and the industry and services, 2014. *Source* Google earth. Elaborated by author

Fig. 3.16 Asphalt roads are shared: market, activities, pedestrians, cars, motorbikes, Abies trees… all in one space. Photos by author

Fig. 3.17 Asphalt roads in afterschool hours. Photos by author

Fig. 3.18 Bus stop on the asphalt roads. Photos by author

The concrete surface (Fig. 3.14, 3)

The village roads and the concreted dikes of *yu* constitute a continuous concrete surface (Fig. 3.19).

Each *yu* has its concrete surface on the peripheral dike, from which several short branches of east–west spaces stretch to the middle of the *yu*. This is because of the specificity of the houses which are built first on the peripheral dikes and always face south. The concrete forecourts (always facing the south) of the peasants' houses visually widen the concrete space of the road. The forecourt is often used as a place for activities such as parking, washing cars, laundry, etc., which aggregates the pollution sent to the field. The vegetation along the roads is random—decorative or productive, indigenous or exotic—and in most cases bold (Figs. 3.20, 3.21, 3.22, and 3.23).

In many *yu* today, two new concrete roads—one from north to south, and the other from east to west—cross the center of the *yu* and connect different *yu*. Thus, in the center of the yu, small industries, shops, temples, and especially new houses are being built—a similar but more domestic situation has happened to the asphalt road. It is gradually changing the morphology of the village—the new houses in the center are larger and more independent of each other, which marginalizes the original houses at the perimeter. The center, which was a void, is becoming denser. Sometimes, an additional layer of asphalt will cover those two concrete road crossings, which privilege cars.

In short, the hard surface can be categorized very roughly into two types: the arterial asphalt space with fir trees, and the capillary bold plain concrete surface that reaches each house. The linear asphalt space is a grid of roughly 10×10 km, and the concrete surface crosses and borders each *yu*. They are both used as public spaces due to the lack of public space in the village, and small and micro-centers are emerging at their crossing points.

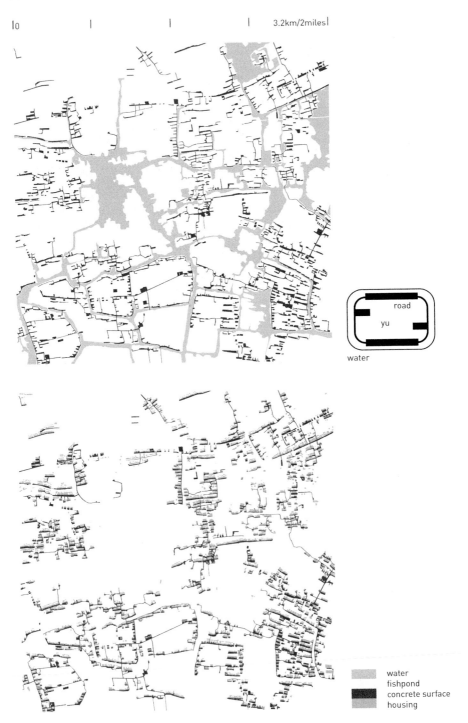

Fig. 3.19 Overlapping of the concrete surfaces with the elements of *yu* and housing, 2014. *Source* Google earth. Elaborated by author

Fig. 3.20 Typical enlarged concrete waterfront oriented east–west. Photos by author

Fig. 3.21 Informal toilets parking space. Photos by author

Fig. 3.22 Narrow north–south concrete waterfront. Photos by author

Fig. 3.23 Car-washing business sending the wastewater to the field. Photos by author

3.2.4 Element 04: Peasants' Houses

Almost all the residential buildings in the countryside of Yangtze River Delta are freestanding buildings or row houses, usually with additional building volumes in the back of the main building (Fig. 3.24). The number of these peasants' houses have been increasing significantly (Fig. 3.25). While residential high-rise buildings are being built next to the field as the cities expand (Fig. 3.26)—a typical image of urbanization in today's China—another massive building process is taking place within the fields: large peasants houses extrude in the middle of the farmland (Fig. 3.27).

industry+service
peasants' houses: main building in the front
peasants' houses: additional building in the back

Fig. 3.24 Different types of buildings in the 3.2 by 3.2 km case study area, 2014. *Source* Google earth, elaborated by author

|0 | | | 3.2km/2miles|

peasants' houses: 2009
peasants' houses: 2009-2013

Fig. 3.25 Increase of peasants' houses in the fields in the 3.2 by 3.2 km case study area, 2009–2013. *Source* Google earth, elaborated by author

A traditional one-floor peasant house in the Yangtze River Delta consists of five parts, each parallel to the other: a front court, a living room, a courtyard, a row of bedrooms, and a back yard with a toilet. Today, land in rural areas is owned collectively by the villagers, and each family owns a small piece of land for building. In Tangqi, the first modern peasant houses were built in the 1980s, buildings of 2 to 3 floors with additional small buildings in the back for storage and activities and an open public forecourt. Recently, 4-and-a-half-floor or even 5-and-a-half-floor houses[5] are

---

[5] According to the Administrative Measures of the Self Constructed Peasants' Houses (2006 and 2015) by Yuhang District, which Tangqi belongs to, a peasant's house must not be higher than 10 m (13 m with a pitch roof), and no more than 3 floors plus an elevated level (no higher than 2.2 m)

Fig. 3.26 City-making process by the field. Photos by author

Fig. 3.27 In-situ construction in the field. Photos by author

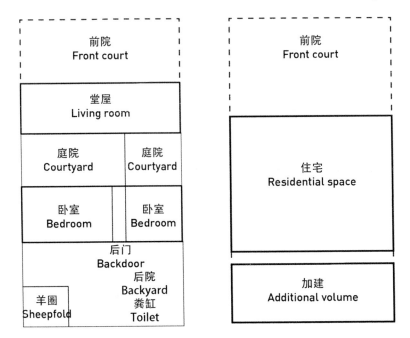

Fig. 3.28 Comparison of the plans of a peasant house in the 1930s (left, based on the diagram by Fei, 1938) and 2000s (right, by author). Elaborated by author

replacing many of the low-rise buildings (Fig. 3.28). These buildings are not purely residential; on the contrary, they are built with space for activities, accommodation for the family of the owner, rent apartments and studios for migrants, etc. It does not only transform the villages quantitatively but completely redefines the atmosphere of the rural area. The high and continuous façade, the elevated main entrances, the monumental staircases, and the large fenced or walled forecourts… all contribute to an urban feeling and signal a new society emerging in the *desakota* with a more complex population, an accumulation of private wealth, and a need for security and privacy.

Peasants' houses from the 1980s and 1990s (Fig. 3.29, 1)

These houses usually consisted of 2-and-a-half- or 3-and-a-half-floors made of brick, with simple tiles as decoration, or simply a layer of cement. All the houses were oriented to the south. There are mainly two types of roofs: a pitched roof as the third floor, or a flat roof with very small pavilions added onto it. They have large windows, loggias, and balconies (usually the length of the façade) on the south side,

---

beneath if necessary. It means a ground floor of 2.2 m, 3 upper floors each 2.6 m, and a 3 m pitch roof. The plot for building, including the affiliated buildings, cannot exceed 125 m^2 for a family with more than 6 members, 115 m^2 for a family with 4–5 members, and 100 m^2 for a family with less than 100 m^2, if the plot is arable. Whether those measures are observed is obviously doubtful.

Fig. 3.29 Photographic inventory of the elements of the housing. Photos by author

with small windows on the north side. The windows are single glazed. Each house has a forecourt to the south without fences, and usually an additional building or a small backcourt to the north. Many of the houses are built in rows, and very few are freestanding.

The ground floor of peasants' houses from the 1980s and 1990s (Fig. 3.29, 2)

The ground floor has a porch the width of the entire building, a transition between inside and outside. A lot of activities happen here daily: cleaning vegetables, drying food, reading newspapers on a chair, washing clothes, meeting neighbors, etc. In front of the porch, it is a large concrete surface for domestic use but completely open to the public, without a fence or walls. A part of the space is kept free as a road, and beyond that, small and affiliated volumes are built. This sequence of space is repeated for each house in the row, which creates a continuous dynamic façade and a linear collective space full of life.

Houses built in the 2000s (Fig. 3.29, 3)

The houses have 5-and-a-half-floors, made of brick and roofed with pitch. They usually have large balconies with massive pillars. The buildings are close to each other due to their volume, which almost gives a feeling of an urban façade.

Usually, the main part of the house has three parts that are disconnected deliberately from each other. The first part is the ground floor, which is for non-residential purposes: shops, warehouses for factories, garages, workshops, etc., and can be rented out; therefore, there is no vertical circulation between the ground floor and the upper floors (Figs. 3.30 and 3.31). The second part is the residential unit for the owner, which is usually the 1st and 2nd floors of the building. An exclusive entrance and staircase are designed for this part. As the photo shows (Fig. 3.29, 3), the forecourt is divided into two: the left half is open to the ground floor which is dedicated to commercial activities, and the right half is gated as an exclusive entrance for the owner's residential unit. The entrance is in the middle of the first floor, with extremely wide doors and large balconies on top of it. The third part is for the "visitors"—the migrants from other parts of China who work in the industries in the village. Usually, a separate entrance and staircase are designed for this part.

The front of the building (Fig. 3.29, 4)

The building encloses a long space with a fence or a wall to create an incredibly large forecourt—a concrete version of an American front lawn. The large forecourt is used for traditional tasks and as multiple parking plots for cars and motorbikes, even trucks. Loquat trees, popular in the region, are planted in the forecourt, and sometimes, tropical trees are grown to illustrate the wealth of the inhabitants.

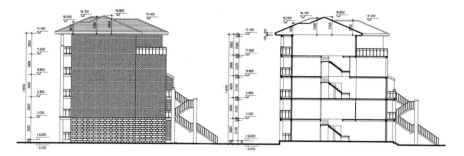

Fig. 3.30 Typical elevation and section of a peasant's house, which shows the internal division in height. CAD drawings collected by the author

Fig. 3.31 House under construction, Tangqi, 2014. Photos by author

Vertical stairways (Fig. 3.29, 5A and 5B)

There are sometimes two patterns of movement. One vertical stairway inside the building is exclusively for the owner who lives on the 1st and 2nd floor. The other stairway is exclusively for the "visitors", the migrants (arrived or expected) who enter the top floors without entering the lower ones. The visitors' stairways are usually external: either at the back of the building (Fig. 3.29, 5A) or at the side of the building (Fig. 3.29, 5B).

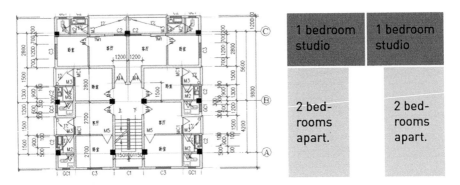

Fig. 3.32 Typical plan for the upper floors of a peasant's house, which shows the spaces of the apartments and studios for upcoming migrants. Elaborated by the author

The top floors are thus designed with a different layout than the part for the owners, which is mainly composed of studios or small apartments with independent kitchen units and toilets (Fig. 3.32).

The back of the building (Fig. 3.29, 6)

Additional construction in the back of the building can be rented by migrants who work for the nearby factories or serve as a small workshop for the family. In this case, it is a small glass factory.

Combined buildings (Fig. 3.29, 7)

Some buildings are combined because of the large family, which produces an even more gigantic mass. The two main first-floor entrances can share a monumental staircase.

The rapidly expanding city-making movement means that massive construction in the field progresses quietly. From 2009 to 2013, the number of houses increased by 4% (from 3940 to 4087 in the study area). With a rough estimation of an average of 3 floors of residential space for each house, the total livable space in the study area amounts to 1,577,400 m^2, which can house 31,548 inhabitants with 50 m^2 per inhabitant. In other words, it means an existing capacity to comfortably house a density of more than 3000/km^2–more than double the current ($\pm$1250/km^2). When looking for a new capacity to house the incoming migrants, we did not realize that a great capacity has been under construction for a long time.

Every architect would be impressed by the flexibility and complexity of today's peasant house, which is designed with hardly any professional involvement. Different typologies are combined within a singular building: in this case, a duplex with a front garden in the lower part of the building, two floors of apartments and studios with a separated staircase not entering the duplex, and penthouse/ apartments with large terraces on the top. The layout of the upper floors is ready for future migrants and can easily be divided or combined according to the size of the household. Occasionally, buildings are constructed with a reinforced concrete frame, which provides even greater flexibility and adaptability.

Countless housing surfaces are built and wasted. A remarkable compensation in the form of cash and/or a "new socialist village" is expected when those peasant houses are demolished during the expansion of the city: the more one builds, the higher the compensation. The huge waste of material, labor, and livable space is not yet considered a design theme.

3.2.5 Element 05: Industry

Tangqi is a typical town with strong agricultural and industrial activities. Industrial Tangqi was born at the end of the nineteenth century. Despite experiencing fast industrial growth and a myriad of industry types, silk was the core industry during the 1920s and the beginning of the 1930s, but it was still far less important than agriculture and handicrafts. Although almost all the industries before 1949 had been destroyed during the war, the number of factories boomed during the Maoist era, especially in the 1950s, 1970s, and 1980s. There were two significant transformations in the Maoist era: the family-based handicrafts were absorbed into the brigades and communes and concentrated and transformed into the modern industry, and industrial development went beyond the borders of the town and pervaded the rural area. A diverse and diffuse industrial base took shape, and industry began to surpass agriculture in value. From the 1990s, industrialization entered another level. In 1998, the 4.9 km^2 Tangqi Industrial Park (Tangqi Machinery Function Zone) was born and attracted large-scale industries. Often overlooked, an incredibly diffuse process of industrialization happened at the same time (Fig. 3.33). From 2005 to 2014, the quantity of all industrial companies grew from about 1000 to 2000, while the upper-scale industries grew from 48 to 109—the expansion of small industries[6] has been as rapid as the upper-scale ones. In terms of value, the small industries contributed nearly 30% of the total industrial output (4.17 billion yuan/14.33 billion yuan) in 2014. This expansion of industry of all scales has led to diffuse consumption of land in the form of family workshops to industrial estates and occupies different spaces in the territory. In the study area, industrial production of all scales covered a land area of 300,394 m^2 in 2013, 48% more than in 2009 (203,442 m^2).

Family workshop inside of the housing building (Fig. 3.34, 1)

A micro-scale metal product workshop on the ground floor of the peasant's house. The equipment is mostly retired machinery from larger factories. The workshop is run by the family, sometimes hiring local laborers, and barely any migrants.

---

[6] Upper-scale industry here means that the annual output of the company is more than 20 million Yuan.

Industrial production 2009
Industrial production 2009-2013

Fig. 3.33 Spaces for industrial production in the 3.2 by 3.2 km case study area, 2009–2013. *Source* Google earth, elaborated by author

Family workshop attached to the housing building (Fig. 3.34, 2)

A micro-scale glass workshop in the back of the peasant's house. The house itself is kept entirely for living.

A small factory (Fig. 3.34, 3)

A micro-scale metal product factory occupies the ground floor and is a simple exten-sion of a peasant's house. The upper floor has been transformed into an office/living space.

Fig. 3.34 Photographic inventory of the elements of industrial production. Photos by author

Fashion factory (Fig. 3.34, 4)

A small-scale fashion factory takes the space of two residential buildings and the space in between. A proper gate has been constructed. Small trucks can park in between the buildings.

Warehouse in the middle of the *yu* (Fig. 3.34, 5)

A small warehouse located in the middle of the *yu*, where two new roads crossing the *yu* meet. Around it, several small factories, a small shop, and a temple have been constructed. Together, they create a micro-center of the *yu*, which had been a void and was the least preferred site historically.

Warehouse along the county road (Fig. 3.34, 6)

A small warehouse is located on the asphalt county road, as part of the façade of continuous activity along the road. It is neighbored by car-washing shops, a clock repair shop, small restaurants, a supermarket, and other activities. The fir trees in front of it have been chopped off.

Factories in the field (Fig. 3.34, 7)

A small-medium metal product factory is located near a large lake with fishponds. The logistics rely entirely on trucks and cars.

Industrial estates (Fig. 3.34, 8)

New industrial estates are developed in large-scale "shoeboxes", surrounded by agriculture. Within the estate, rows of mid-rise housing have been built for future laborers. Many empty plots, waiting for new companies to occupy them, are used as vast vegetable fields by the inhabitants.

The essence of *desakota*—the mixture of agricultural and non-agricultural activities—is becoming ambiguous. Industrialization is complete, and the role of agriculture is more and more marginal. In 2014, the agricultural output of Tangqi was 0.57 billion yuan, 4% of industrial output. This can be explained with a micro-story about the textile industry in Tangqi.

Micro-story of the textile industry

The success of the textile industry can be largely attributed to a flexible network of production, varying tremendously in scale: several large-scale enterprises in the industrial estate, small and medium-sized factories in the villages, and micro-workshops in the peasants' houses (Figs. 3.35 and 3.36). The latter two especially form a dense and extensive network in the field.

Fig. 3.35 Small family workshop of textiles. Photo by author

Fig. 3.36 Storage space, the office, and the production are all in one space. Photo by author

Yunhui village has a population of 2543 distributed over 20 natural villages. There are more than 100 textile enterprises with a total of around 400 spinning machines—4 spinning machines per enterprise, a perfect size for a family workshop. The enterprise of Mr. Shen, a small factory of no more than 200 square meters with 10 employees and 24 spinning machines, is already among the top 10% of enterprises in the village in terms of size (Fig. 3.37). Every employee is a local, except for a few migrants

Fig. 3.37 Factory of Mr. Shen, a textile factory in the field. Photo by author

Fig. 3.38 Yundi factory is the biggest textile factory in the village. Photo by author

who are married to locals. The biggest enterprise "Yundi" (Fig. 3.38) employs 60 people with more than 100 spinning machines, while the smallest enterprise is run by a couple in their own house, with up to 6 spinning machines. 60% of the village's textile production is made by such family workshops.

The relationship between the relatively small and large enterprises (although they are small in general) is a network in which both competition and collaboration coexist. The clients are from both China and abroad, especially from the Middle East like Dubai, through the textile trading center located in Keqiao—a town in the Yangtze River Delta. Some of the enterprises have market representatives in the trading center, but most are too small and work only through outsourcing orders from Keqiao or via other enterprises in the same or neighboring villages. Mr. Shen's enterprise has two production challenges: it does not have the most updated equipment, which adds an unaffordable extra cost to high-level production when the output is too small; the quantity of orders varies widely: from 200 to 800 pieces, making it difficult to deliver on time. The solution is to outsource to the neighbors: 40% of the orders are outsourced, to not only the smaller but also the bigger enterprises. At the same time, Mr. Shen's company also receives orders from other companies. The biggest enterprise in the village even has three specialists who deal with their outsourced enterprises in Zhejiang, Guangdong, and Jiangsu.

The flexibility of the production, which reduces the cost significantly, lies in mainly three elements

The first element is the scale or the very nature of the small, especially tiny enterprises themselves, i.e., the low capital investment, the flexible working times, and the shared management network. The capital invested in the land and the industrial building is almost negligible—a workshop of 30 square meters in the backyard. The outdated machinery, usually second-hand and more than 20 years old, still works due to a large number of orders for low-class products. The motivation to produce for oneself or one's own family is so high that a couple could work 24 h a day in shifts when production requires it; and since they are at home, they can at the same time take care of their children or grandchildren when production permits—a much more flexible calendar than a fixed contract with an employee. In this system, one is at the same time the worker but also the accountant, the seller, the driver, and the maintenance man—sometimes for yourself, sometimes for others. This network saves on all types of "departments" needed in a modern enterprise and resembles the utopian territorial industry concept by Okhitovich (1930): there are no centers and no clear borders to each factory. It is more like a network of different workshops, departments, laborers… a flexible entity.

The second element is the coexistence of industrial and agricultural production. The countryside provides spaces for small factories with few costs: the 50–200 square meters can be housed easily in the backyard of any peasant's house. The whole village has 646 households, each has an average of 5.4 *mu* (or 0.36 ha) of agricultural land (not so different from Fei's Kaixiangong village in the 1930s), and 2.9 *mu* (or 0.19 ha) of open water surface; thus, the fishponds are very much developed in this area. Textile production, for example in Mr. Shen's family, is one type of activity among many others. The family also runs a fish food factory and a cold storage warehouse, both serving local agriculture. Mr. Shen was assigned to the fish food industry during a decline in the real estate market, resulting in a shrinking need for textiles, and

Fig. 3.39 Factory owner Yang explains their experience of being a farmer before working in industry. Photo by author

sent back to manage the textile business when the textile market recovered. When the fieldwork draws intensive labor on busy days, workers get a vacation from the factories. However, this mutual resilience is becoming rare. In some parts of the territory, the agricultural sector can contribute around 10% of the family income, and in others, it is rather marginal—it is a fully industrialized countryside (Fig. 3.40).

The third element is the very type of labor, which is in crisis in today's urbanization. Most of the owners of the micro-workshops are in their fifties and were born and raised in the Maoist era in a peasant house and are familiar with fieldwork. When they were in their thirties and the rural industries started to boom, they worked in and got familiar with industrial production. Now, they run their own business, motivated by the money but not relying on it: their children are grown-up and earn for the family, and the absence of financial pressures enables them to accept a lower price and to quit any time they want. If urbanization, in the form of urban concentration and rural degradation, continues, the current owners will not have their successors in this family business (Fig. 3.39).

Of course, this model of "mutual aid", as Kropotkin (1901) would say, cannot be applied to all types of industry. The heavy machinery equipment industry also exists in this territory but is more concentrated in industrial estates without any spin-offs in the villages due to its tech-intensive nature and large spatial requirements. The metal products (for example, steel cables) industry also has various scales of enterprises in the industrial estates/villages, but it is mostly characterized by competition due to the homogeneity of the product, and the immense pollution those family workshops create. The textile industry does not create severe pollution in water, air, and noise. The cost of transportation is low due to the well-developed network of roads in the countryside, the network of small-scale transportation using trailers (another example of agricultural-industrial mutual aid), and the local logistic services provided by some factories. The concentration in industrial platforms under these conditions does not

Fig. 3.40 Manager Liang explains his work both in the factory and in the field. Agriculture produces more than 10% of the family income. Photo by author

provide an evident advantage; on the contrary, the dispersed small enterprises in the field nourish the whole industry.

A new *desakota*

The example of Tangqi proves that the third space today still presents some of the features of a *desakota*: the increase in non-agricultural activities, the extreme fluidity and mobility of the population, the intense mixture of land use, the increased participation of women in non-agricultural labor, and being "invisible" or "gray" from the viewpoint of the state authorities (McGee 1991, p. 16). But there are fundamental ongoing changes to the essence of the *desakota*: the relation between agricultural and non-agricultural activities.

The industry has replaced the paddy field and become the fundamental reality of the "third space". The inhabitants are pure workers or worker-farmers instead of farmer-workers. If according to McGee, the more urban condition means more than half of the labor and output are from non-agricultural sectors (McGee 1991, p. 21), the third space is far beyond it (higher than 90% in both indicators). More significantly, since the end of the twentieth century, the industry has detached itself from agriculture. The orchards, sericulture, and rice determined the industry in Tangqi based on food processing for most of the twentieth century. Since the 90s, most of the raw materials are imported even though the fruit industry is still important in Tangqi (Dong 2014: 60). Today, the four cornerstones of Tangqi's industry are chemicals, food processing, metal products, and textiles and fashion, none of which rely on local raw materials. The domination of the silk industry is over.

There is a diffuse network of industry and urbanity in the field. Agriculture has become to some extent more of a life support system, and a background landscape, providing ecological and environmental values. The agricultural territory, which can feed its industry, according to Kropotkin, Wright, Okhitovich, and many others, has turned out to be a fallacy in the third space in the Yangtze River Delta. What they imagine is the strong competitiveness of this diffuse system at all levels of industry, which will be strengthened by new technology such as the smart grid. Today's factories are already starting to be connected through the Internet, to share orders, resources, space, materials...This is going to be a much more advanced and efficient version of today's network based on human relations.

However, the third space as a legitimate space for absorbing the huge incoming population has not been considered by the current policy—it is not fully recognized as an urbanity. Its potential to do so is being compromised by the unclear future of the existing industrial system and the decline of agriculture, both as an economic sector and an environment to live in.

3.2.6 Element 06: The Elementary School

The elementary school is studied as an example of the school system, and an example of the organization of an environment for facilities in the third space. The facilities in the rural area, which are supported by a rural infrastructure and serve a relatively rural community, need to be adapted to their increasingly urban surroundings. The declining performance of some of these facilities, because of demographic change and the economic reality, is used as an argument for their more centralized redistribution. There are three schools in the 3.2 km by 3.2 km study area: the Hongpan Central Elementary School, its Weijiadai Campus, and the Chaijiawu Campus of Chaoshan Central Elementary School. The first two schools, managed as one school, are used here as a case study (Fig. 3.41).

Micro-story of the Hongpan Central Elementary School

In 1935, the Mojiaqiao School, the first elementary school in this area, was established with 3 classes and 75 students. During World War II, it was extended to 4 classes with 120 students.

From 1949, the idea of the school was to "build the school at the door of the inhabitants". The school system was decentralized. Besides Mojiaqiao School, Weijiadai School, and Tangbingdun School, which were categorized as "complete schools", schools with the full 6 grades, each village had its own village school.[7] A village school did not have the complete six grades usually and was not able to issue graduation certificates to its students. The students at the village school would spend a few years in the complete schools to finish their studies. Hexidai School was one of those village schools. It had 30 students from seven to twelve years old and shared its

---

[7] In principle, the schools that are not central schools, both complete and incomplete, are categorized as village schools. But among the inhabitants, the small and/or the incomplete schools are usually called village schools.

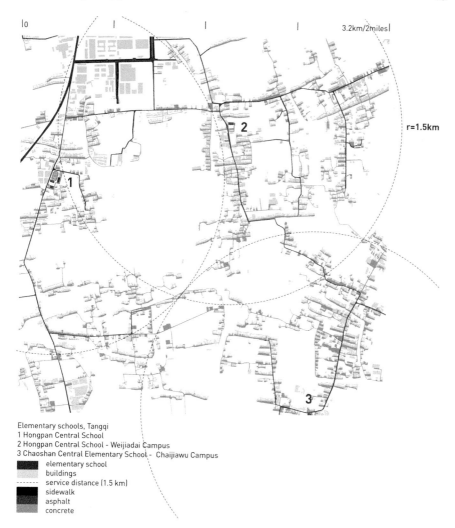

Fig. 3.41 Elementary schools in the 3.2 by 3.2 km case study area, 2014. *Source* Google earth, elaborated by author

building with the production brigade. The complex became an administrative office for the brigade, the food production center, teacher accommodation, the village auditorium, and classroom. The students followed classes in the morning and went back home for agricultural work in the afternoon and sometimes received military trainings in the school (Fig. 3.42). All the activities were carried out in the same courtyard, the playground of the school. In 1963, Hexidai School started to operate as a "耕

Fig. 3.42 Military training of the students in the Hongpan Central School in the 1970s. *Source* The Hongpan Central School

读小学" (cultivate-and-read school) in the evening, to educate illiterate youngsters who worked during the day.[8]

This decentralized system persisted after the Maoist era. Until 1999, there was one central school, two complete schools, and seven village schools in this area, accommodating 1157 students within 31 classes.

In 1999, the school system was centralized. All the village schools were decommissioned, and the students were integrated into three complete schools. In 2003, only two complete schools remained: the Hongpan Central Elementary School, and its Weijiangdai Campus. The Hongpan Central Elementary School covers three villages and has 12 classes and 517 students in 6 grades, its Weijiadai Campus covers two villages and has 443 students in 6 grades and 11 classes.

According to the interview with Mr. Wei, the principal of the Hongpan Central Elementary School, the school (including its Weijiadai Campus) serves five villages and approximately 30,000 inhabitants. Half of these are locals, and the rest are migrants. The current policy allows the children of migrants to enter the local school if they have a residence permit valid for one year, and they have paid social security for one year. 1/3 of the students are from migrant families. The number of children from local families could have filled the school, but 1/3 of them chose to study in schools in the city or the town, where the quality of education is perceived to be

[8] The evening elementary school, a combination of production techniques and education, was not a solitary event. In the 1960s and 1970s, the Tangqi High School opened Xuenong (学农, literally meaning "learning from agriculture") campuses in several production brigades and villages. In 1984, the first class on freshwater aquaculture was set up in the school.

Fig. 3.43 Table tennis class in the arena of the school. *Source* The Hongpan Central School

Fig. 3.44 Outdoor training of table tennis. *Source* The Hongpan Central School

better. No students live farther than 1.5 km from the school, less than 20 min on foot. All of them are sent to and picked up from school by their parents, mostly by car.

In the Yuhang District of Hangzhou, where the town of Tangqi is located, there are 45 central schools. Hongpan Central Elementary School ranks 37th. The school invested significantly on sport and especially table tennis (Figs. 3.43 and 3.44), but it is still struggling to get the best students and teachers. Although its sports education ranks 25th and its table tennis 1st, it struggles to stop the outflow of the best students toward the nearby city and town and the best teachers' unwillingness to come.[9]

---

[9] New teachers are allocated to schools annually. The best new teachers have the right to choose the schools. Therefore, it is difficult for lower-ranked schools to obtain good teachers.

The Hongpan Central Elementary School is expanding, from 12 to 24 classes, 18 of which will be used immediately after construction. The expansion will cost around 20 million Chinese Yuan (2.5 million Euro) and consume 0.4 ha of water field. A new kindergarten of 0.75 ha will be built next to the expansion area. The motivation for the expansion is the obsolete equipment (according to Principal Wei) and the rapid growth of the population (according to the vice-mayor of Tangqi responsible for education).

The rural schools: from decentralization to centralization

One of the main debates on the rural school system in China is about the "decommission of the educational stations and the amalgamation of the schools" (撤点并校); in other words, whether the diffuse and decentralized system should be adjusted toward a more centralized system. In the Maoist era, the main idea of the elementary school system in the rural area was for "each village to have its own schools" (村村办学). The system comprises the central schools, the complete village schools, the incomplete village schools, and the educational stations. A central school usually serves a township-level administrative unit. The village schools are affiliated with the central schools. Some of the village schools have the complete 6 grades and are called complete schools; the others have not all the 6 grades and are called incomplete schools. Their students must spend some grades in the central or complete schools. In very remote areas, such as in the mountains, educational stations have been built, each of which might have only 10 students or less. From the beginning of the 1980s, especially after compulsory education was implemented in 1986, the first wave of decommissioning and merging of schools in rural areas occurred. From the end of the 1990s, but especially from 2001,[10] the second wave of redistributing the elementary and high schools in the rural areas took place. The reason why this centralization process was necessary was stated clearly in the 2012 paper "Suggestions on Regulating and Adjustment of the Distribution of the Rural Compulsory Education Schools" by the State Council: the growing number of children migrating into cities with their parents, who started to work in the cities, and the lowering birth rates in rural areas. The objective was to adjust the distribution of the schools to the geographic change of the school-age children (Shan and Wang 2015). From 1998 to 2016, the number of elementary schools reduced by 70.9%, and the number of students by 29.0%,[11] most of which occurred in the countryside. There are positive aspects to this process, especially in terms of quality improvement of the education and the schools. However, there are also negative aspects including the exclusion

---

[10] A sequence of relevant policies was implemented, including "The Decision on the Reform and Development of the Basic Education" by the State Council of the People's Republic of China in 2001, "The Notice on Improving the Management System of Compulsory Education in Rural Areas" by the State Council in 2002, and "The Management of the Special Funds for the Adjustment of the Distribution of Primary and Secondary Schools" by the Ministry of Finance of the People's Republic of China in 2003.

[11] Source: The Annual Statistics on National Education Development, published by The Ministry of Education of the People's Republic of China.

of students who live in remote and poor conditions, the increasing financial and time costs of the boarding schools, and longer commuting distances, as a result of over-centralization which are a consequence of financial concerns of local governments.[12,13] The decentralization-or-centralization debate hides the deeper issue of choice between the quality improvement of education in certain schools, and the underlying problem of the increase in the differentiation between different school levels with a more balanced development of the schools. While the central school in a town benefits from the concentration of resources and the decommissioning of the village schools, the difference in quality between it and an upper-level school, for instance, a city school, is increased by the same logic. The "decommission[ing] of the educational stations and the merg[ing] of the schools" cannot bring about a balanced development (Shan and Wang 2015).

The schools in Tangqi (especially in the study area), as schools in the third space, have also undergone a centralization process. However, because the population is growing instead of reducing as in the "typical" rural areas, this process faces resistance. It is not the typical resistance from the children left behind in the remote low-density rural area as a result of the concentration of schools. It is a resistance from the inhabitants who would like to have a school nearby, not only for convenience but also as a central point related to the gathering of people, the success of enterprises around it, and even to the housing prices. At the same time, the schools have unique educational and spatial difficulties. They cannot catch up with the quality of the schools in the town and city nearby through concentration. They are rural schools but are receiving an increasing number of migrant students. After the disappearance of many village schools, more space is needed today, which would consume a large area of agricultural land. While great attention has been paid to the schools in remote rural areas, the schools in the third space rarely appear in the debate and research.

The space of the school

The Hongpan Central School is expanding and transforming into an urban-scale school that has a significant impact on the space of its rural surroundings. It is starting to create a mobility and safety problem as stated in the official introduction document of the school. The main gate of the school is located along the bus line. Due to the narrow road, the kindergarten located on the other side of the road, and the supermarket and the food market nearby, the safety of the commuting students is at risk.

---

[12] See also a detailed case study of a typical town in mid-western China by Shan and Wang (Shan and Wang 2015).

[13] To correct the unnecessary decommissioning and the over-concentration of the schools, "The Advice on Regulating the Adjustment of Rural Compulsory Education Schools by The State Council's General Office" was published in 2012. It states that the distribution of schools should ensure that students receive education nearby, for instance, boarding school is not allowed for grade 1–3 children.

The front gate of the Hongpan Central Elementary School (Fig. 3.45, 1A and 1B).

The Hongpan Central Elementary School has 517 students today. The front gate is located on an asphalt road, with trees on both sides. Due to the expansion of the road to accommodate the increasing number of cars and public transport, there is little space for trees, and many trees have been cut down. There is no space for sidewalks. On the other side of the road, there is a row of shops. The space in front of the shops is dedicated to vendors and logistics. To regulate the logistics vehicles and prevent them from parking on the road, fences at the edge of the asphalt road surface have sometimes been erected. As a result, there is no space left for the students who come out of the school, which puts them at risk. The school itself is completely fenced off from its surroundings.

The students being picked up at the Weijiadai Campus (Fig. 3.45, 2)

The Weijiadai Campus has 443 students today. After school, the parents, the students, and different types of vehicles block the entire road. About 200 m to the north, there is a bus stop, and about 100 m to the south, there is the village food market. The school and the market bring in a great number of buses, customers, and logistic vehicles.

The Square of Culture of Weijiadai (Fig. 3.45, 3)

On the other side of the wall of the Weijiadai Campus and the kindergarten next to it, it is a large lawn, a space without a clear purpose and under-used, a space that is meant to be public but is disconnected from the public facilities next to it. At the same time, the campus needs space inside and outside the school.

The main classroom building of Hongpan Central Elementary School (Fig. 3.45, 4)

The 3-floor building was built in 1992 and is out of date by today's standards (for instance, the fire regulation). The space between the main gate and the building, which lies around a small but beautiful green garden, is used today as parking for the staff of the school—the students must pass between the cars.

The sports field and the arena of Hongpan Central Elementary School (Fig. 3.45, 5)

Hongpan Central Elementary School has an open-air sports field of more than 6000 m², including a 200 m running track, a small soccer field, a basketball field, and a volleyball field. It also has a cultural and sports arena, with an indoor basketball field, table tennis zones, and music and dance classrooms. Its Weijiadai Campus, although with a similar number of students, has an open-air space of only 500 m², with one basketball court. All the physical education in Weijiadai Campus takes place within its own space, and the students do not use the space on the central campus after school. The Olympic games of the school take place on the central campus, and the students go there on foot accompanied by the teacher. There are currently two

Fig. 3.45 Photographic inventory of the elementary schools. Photos by author

girls from the Weijiadai campus being trained in the central campus for the dance competition. The teacher drives them by car (1.7 km) to the other campus.

The government has requested that the sports fields and the arena are open to the villagers on weekends and school holidays. But the school is reluctant and resistant to do this.

The water field of Hongpan Central Elementary School (Fig. 3.45, 6)

At the back of the school is a water field that is still in operation. The field is enclosed by a tall concrete wall, while all walls given out on the streets are transparent. From the dance classroom in the arena, one can only see a blind wall and not the field beyond. In future, this water field will disappear due to the expansion of the school.

Today, there are no lessons in the field or the factory. All the educational activities are conducted within the campuses

Streams and water banks (Fig. 3.45, 7)

Around the school, natural features can be found in small streams and trees on the banks of the ponds. One can imagine and design a beautiful public space. At the same time, these spaces are abandoned today and seen as the back or left-over spaces behind and between the peasants' houses.

The expansion of the school

The expansion includes the new part of the Hongpan Central School. It will have a surface of 4224 m^2, with two four-floor buildings that accommodate 24 standard classrooms. When the new buildings are ready, 18 classrooms will be in operation immediately, and the other 6 will be used later. There will be 38 parking spaces on the ground floor level. The Hongpan Second Kindergarten will also be extended to 7514 m^2, with one three-floor building and one four-floor building with 15 standard classrooms. There will be 32 parking spaces on the ground floor.

From the interview with Principal Wei of the Hongpan Central School, the future of the two campuses is uncertain, which is why he decided to build a new part with spare capacity.

The schools in the third space

While the schools in the urban area are struggling to have enough classrooms and playgrounds for their huge number of students, the schools in the "typical" rural area are struggling to have enough students and consequently enough funds to maintain their quality and even existence, the schools in the third space—that is still categorized as "rural"—have their own particular challenges. Their main challenge is how to fit the expanding school into the spaces that were planned and constructed for a rural area when this rural area is gradually transformed into an urban one; how the school can fit in a changing context in terms of population density, traffic, and activities intensity, and the new pattern for the use of space.

When the schools were built, the safety of commuting students, the parking, and the capacity of the buildings was not considered problems. Today, the future of schools is largely influenced by the complexity and uncertainty of their surroundings: the capacity of the road, the number of incoming migrants, the intensity of the logistics and other activities, the reduction of the agricultural field, the design of the civic space around it, etc. A school has become also a more complex object, whose location and scale have a considerable impact on the housing prices around it, whose quality is to an extent determined by its urban competitors, and whose consumption of agricultural fields is compensated for by other counties in the delta.

At the same time, schools are still situated in a rural setting. This rural setting has its physical aspect, as a great portion of the territory is still agricultural. The streams, orchards, and water ponds provide not just spare spaces for the schools to expand, but also a unique environment, which has the potential for the creation of a different type of school. The proximity to different types of agricultural and industrial production spaces offers the potential to operate a different type of education. The rural setting also has a social aspect. It is a society that is gradually transforming from rural to urban in this in-situ urbanization process. The villagers are searching for a cultural square, for new types of civic space; they are demanding an arena, cinema, auditorium, sports fields, library, and other facilities. They are leading a more urban and even metropolitan life. The future of the school is contextualized by this transformation of the rural setting, and the future of schools could also intervene in this transformation and create new relations with this setting.

3.3 An Urban–Rural Division by Elements

On one hand, in many ways, the 3.2 km by 3.2 km area of Tangqi is already an urban area—not only in statistical terms such as population density, income, and activities but also in physical terms of roads, distribution of services, housing typology, etc. (Fig. 3.46). However, the quality of the physical environment tremendously compromises livability compared to the urban environment (Figs. 3.47 and 3.48). The urban–rural division is not only presented by the standard in income, services, infrastructure, energy supply, etc., but also by the quality of space and by its articulation. As Bernardo Secchi claims, the "great articulation of the spaces" the historical city is precisely what distinguishing it from the periphery where the "dramatic" reduction of the articulation of the space takes place. (Secchi 1991).

This notion is shared by Friedmann, who imagines "transform[ing] the countryside by introducing and adapting elements of urbanism to specific rural setting[s]" (Friedmann and Douglass 1978: 182), he has in mind the space in rural areas in Europe. The "articulated" and "urban" elements can be seen in many European rural areas. A simple rural street of 2 lanes could be a complex element composed of an asphalt surface with gutters on both sides, a line of bushes to separate the cars

|0 | | | | 50km|

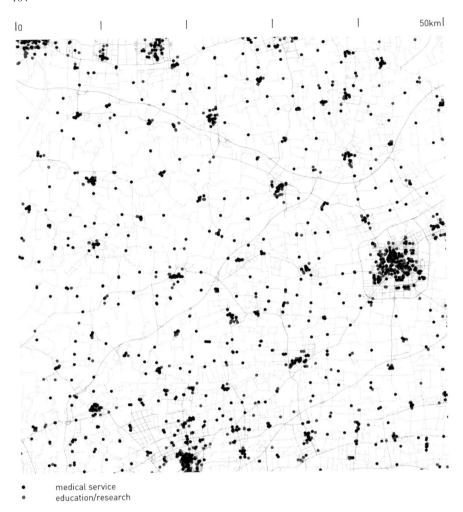

•   medical service
•   education/research

Fig. 3.46 Public facilities in a 50 × 50 km frame of the Yangtze River Delta, between the town of Tangqi, Tongxiang, and the city of Huzhou. *Source* GIS data, 2010, National Earth System Science Data Sharing Infrastructure China. Elaborated by Author

from the pedestrians and bikers, a wide concrete lane for the bikes (colored as the one in the urban area, but shared with pedestrians), a natural linear *wadi* in the field, and two lines of trees on both sides of the field: in the village, the road could be designed as a *woonerf*—an urban typology of road, a universal brick surface, which is a similar material to the houses along it, a shared space for cars, pedestrians, and bikes, with a specific design of the space by inserting elements such as trees, flowerbeds, and parking lots to slow down the traffic and create a domestic atmosphere. Social housing—which is usual in urban topology in China—can be found in low-rise buildings along the field, with a piece of lawn fenced lightly and

Fig. 3.47 Lack of articulation: a public playground in Tangqi. Photo by author

Fig. 3.48 Lack of articulation: a public space in-between houses in Tangqi. Photo by author

Fig. 3.49 Photographic comparison of the public space in rural China (Tangqi) and rural Europe (Stabroek, Belgium). The photos of Tangqi are labeled with A, and the photos of Stabroek are labeled with B. The comparison doesn't attempt to show the distinction in the quality of public space between two completely different contexts but attempts to show the possibility to articulate the public space by utilizing urban typologies of space in the rural area. 1—The road passes in the middle of fields. 2—The space on the side of the neighborhood. 3—A road in-between buildings. 4—The relation between houses and landscape. Photos by the author

transparently, as part of the village green.[14] In general, many of the urban elements are also implemented in rural areas, not always with the same appearance but with the same level of articulation (Fig. 3.49).

The lack of articulation reflects an urban–rural divide in the idea of space. Take public space as an example: the notion and form of public space in China have evolved dramatically through the different phases of urbanization. The urban public space and the rural public space went through two different trajectories and evoked completely different reflections in current studies, almost as if they were entirely different objects.

3.3.1 The Urban Public Spaces

The premodern Chinese cities are made of walls, without public squares equivalent to the western agora or forum. The Chang'an city in the Tang dynasty was made in a grid of large plots enclosed by pure walls and occasional doors forming the entire façade of the streets. The two markets—the Western market and the Eastern market—are both walled spaces. Although by the end of the Tang dynasty, the walls were being removed and commercial streets started to appear, the wall remains a symbolic element to close off a large space for crowds for specific functions (royal palace and garden, temples, markets, etc.). In the second half of the seventeenth century, the first Catholic cathedral was erected in Beijing: the Cathedral of the Immaculate Conception, which took the form of a courtyard as its main outdoor religious space instead of an open forecourt. In the nineteenth century, when the modern idea of public space reached the east coast "treaty ports" in China, the new public parks were walled and gated (Gaubatz 2008: 74).

Before the introduction of the "week" in China, the use of public space was inseparable from the rhythm of work and traditional events, both strongly related to the seasons. The temples usually enclosing open and natural spaces and relatively accessible to everybody functioned as important public spaces (Gaubatz 2008:74). The three main groups of public activities—pilgrimage, viewing (usually in seasonal flows), and festivals—taking place at specific times of the year and in specific temple gardens in Beijing are well explained in Ju's study (Ju 2016). The same activities, especially the pilgrimage and the festivals, also took place in rural areas with almost identical topics and formats and will be explained in later paragraphs.

At the end of the nineteenth century, the concept of public parks was introduced to China when the monarchy was ending and modernization started. It was officially promoted by the city administration of Beijing in 1914 after the implementation of the "week" from 1902 to 1911 (Ju 2016, p. 54). The idea behind it was a Haussmannian modernization of the city with large open streets and spaces that would facilitate a healthy urban life. While public parks were created from old royal gardens, the poor

---

[14] The example used here is Stabroek, a municipality on the outskirts of Antwerp, Belgium. It is in a rural area, with a population around 18,000 and a density of 860/km^2.

could not enjoy them due to the high entry fee and the traditional calendar of work they still practiced unlike the elites, and also to the fact that the monumental and seasonal traditions they were familiar with were not meant to be practiced on weekly bases. The traditional public space and its use were shrinking and oppressed due to the expansion of the new public spaces (Ju 2016, p. 54).

In 1949, the western "square" started to be implemented in China, symbolized by the demolition of the house in front of Tiananmen and the creation of Tiananmen Square. On the one hand, wide streets, squares, and parks were built; on the other hand, the cities "grew as accretions of these walled and gated cells, and many of these walled complexes containing not only spaces of work but also residential, educational, recreational, and social service spaces, public life retreated, to a large extent, to the confines of the home compound" (Gaubatz 2008, p. 75; Gaubatz et al. 1995).

After the Economic Reform of 1978, the market economy demanded different types and forms of public space. Shopping malls and gated residential communities proliferated rapidly in the cities, while the quality of daily-use public spaces and amenities declined. As a result, the public spaces relied upon by average Shanghai urbanites are much plainer: sidewalks to nearby markets, pocket parks close to their apartments, and a few indoor facilities sponsored by the state. It is surprising to find that, for many crowded existing neighborhoods, urban redevelopments have brought about the least improvement to these "bread-and-butter" urban amenities. In some areas, the situation has even deteriorated.

The sequence of transformation of the society—from the imperial, to socialism, to "Socialism with Chinese characteristics" (combined with the market economy)—determines the types and forms of public space in the city.[15] Given the predomination of the market economy today, a consequence of Deng's "development as the absolute principle" ideology, the "bread-and-butter" public spaces are in crisis.

3.3.2 The Rural Public Space

The main traditional public activities in rural areas are (periodical) trading, seasonal and religious festivals, and weddings and funerals. Correspondingly, the main public spaces in the villages are usually the temple and its *placettes* (small squares), the ancestral hall, and the stage for traditional operas. There are also fields, roads, water wells, dams, and other spaces for daily and informal public uses. The combination of those specific public activities and public spaces is attributed to the shared religion, ways of production, common values and habits, and social groups beyond the scale of a family related to those activities. In Fei's famous case study of Kaixiangong

---

[15] Gaubatz wrote a fantastic narrative of the transformation of the Wangfuji shopping street, from an ordinary neighborhood in Beijing, to a "modern" shopping street at the beginning of the twentieth century after the early influence and connection with the western world from the late nineteenth century, the birth of the "Dong-An" open-air bazaar, a selection of state-run stores and bureaus after 1949, to new shopping plazas after the 1990s (Gaubatz 2008, p. 78). It is a common story repeated in many cities in China.

village, an ordinary and representative village in the southern Yangtze River Delta, there are eleven "*Duans*". Each has approximately thirty families and its particular "King Liu"—a symbol in the Dao religion. Each year, a family from the *Duan* must organize two big events and feasts for the entire *Duan* (Fei 1939, p. 98). "*Zu*", a group of approximately eight families sharing the same family name, organizes the wedding ceremony. The wedding ceremony defines the members of the *Zu*, the relation between whom is looser than a family but much stronger than the clan. For example, the *Zu* can be a mutual aid group at the same time, taking the wedding as an occasion to discuss funding and lending (Fei 1939, p. 85).

During the Maoist commune period, the production brigades/teams were built on the level of a village, to promote singular common ownership, establish concentrated mobilization, and launch the socialist education movement. Such a motivation was presented spatially: from the typology of buildings to the choice of trees in the public space (Fig. 3.50). The brigade/team headquarters and the "grand halls" were the main formal public spaces in the village for meetings, communist festivals, and education and propaganda on the policies. It could be a simple new building but often consisted of transformed temples or ancestral halls—a symbolic transformation of space reflecting the change in social organization. Besides that, there were also public amenities such as the supply and marketing cooperatives, medical services, and the (open-air) cinemas. But in general, the public activities, monotonous in their content

Fig. 3.50 Monumental trees along small roads planted in the Maoist era. Today, the road is used as a public space in the village. Photo by author

and form, were strongly related to the political agenda, not to mention the very limited time after the long collective working hours.

After the Economic Reform in 1978, towns replaced the communes, and the villages and villagers' groups replaced the brigades/teams. The family became the fundamental unit of agricultural production. A vacuum in the public realm, both in society and space, appeared: the room for traditional events, religious activities, and the grand weddings/funerals, which had been occupied by the communist functions, was now available again; but, due to the communist movement and modernization in general, those traditions were discouraged and could not be revived to their full extent. Along with the "city-making" movement that has drawn young laborers to the cities, several waves of self-construction of houses started in the 1980s and broke the traditional concentrated settlements of the villages and expanded them to their peripheries. Since development is the absolute principle, the physical public spaces were recycled to become productive spaces. The symbolic grand halls were rented out for factories, residential spaces, storage, and barns; the grain-drying ground was cut up and redistributed among the villages; the wide public roads were narrowed by the extension of the walled forecourt of each peasant's house. Since the 1990s, the state has tried to enrich the cultural and public activities in the countryside through a "culture down to the villages" movement—a top-down movement of sending books, movies, and other media to the countryside but this is not appreciated by the locals.

Confronted with this challenge brought by the transformation of rural society, from the 2010s, a movement for the construction of "cultural halls" has been promoted by the government.[16] The idea is to refill the "vacuum" with traditional culture: the celebration of filial piety and the ancestors, rebuilding of the halls of ancestors, the teaching of the traditional arts (calligraphy and painting), the construction of "halls of morality", the relaunch of temple fairs, etc., accompanied by the construction of modern amenities such as sports fields and public squares. A trajectory of the transformation of public space along with the transformation of society can be traced. Lu and Cheng present the transformation of public space in Zhejiang villages throughout history in their study: in the Suoyuan village, the ancestral hall of Yan was used for its "Zu" before 1949, housing family events, traditional opera, and education. During the Maoist era, it was used as a granary, a grand hall, and an elementary school for the production team while from 1978 to the 2000s, it was rented out for factories and offices. Since 2010, it was transformed into a cultural hall of the village.

The demands put on public space brought about by the transformation of society have been answered with highly ideological solutions. While public space and the public realm in the urban area have been surrendered to the market economy, the absence of the two in rural areas is today being filled by a recelebration of traditional Chinese values. Especially, the raising of families (or the atomization), the diffusion of the settlement and thus the diffusion of society, and the modernization of the way of life are creating an "urbanity"—or simply a modern society—in the rural area, which could not be satisfied by traditional culture. An ancestral hall today does not

---

[16] In 2013, the Governor of Zhejiang Province made a call for "constructing 1000 cultural halls" in the villages of Zhejiang Province.

mean the same thing, given the weakening of *Zu* in rural society. While the dualism of urban and rural is one of the most serious problems in today's China, where in many rural areas especially in coastal China, rurality is simply a diffused but also dense urbanity combined with agriculture production—a *desakota* or a city in the field. Life there, at least in terms of public activities, is imagined completely differently from urban, or even modern, life. Those imagined public activities can be practiced only in certain exceptional spaces and on particular occasions, showing a lack of concern for the majority of daily public spaces.

Similar divisions are true for other elements in the urban and rural areas. At the same time, society is generating new values and culture, for example, sustainable living—a shift to a renewable energy-based, reuse/recycling economy with a diversified transport system. Light mobility, organic food, urban farming, recycling, and the sharing economy are becoming increasingly popular. In 2014, the National New Urbanization Plan 2014–2020 was released, which clearly stated that all cities should realize sustainable development and greatly improve public amenities. In 2015, the Action Plan on The Adaptation of Cities to Climate Change was published as part of the national strategy on climate change. All those new actions and challenges bring new agendas to public space in both urban and rural areas.

The difference between urban and rural areas is disappearing gradually. When the new values and challenges are ever more shared by the two, a common agenda for their construction through elements could be imagined that doesn't just provide the same quality and functions in both environments, but also physically connects those two as unity, a space in which the inhabitants can freely move and live—a common space.

References

Dong J (2014) 塘栖——一个江南市镇的社会经济变迁. East China Normal University Press, Shanghai

Edwards P (1982) Lecture notes on utilization of animal and plant wastes. Report of consultancy at the Regional Lead Center in China for Integrated Fish Farming. Network of Aquaculture Centres in Asia (NACA), FAO Field Document, no. NACA/WP/82/6, Bangkok, Thailand, October 1982, 104 pp

FAO (1983) Freshwater aquaculture development in China. Report of the FAO/UNDP study tour organized for French-speaking African countries. 22 April–20 May 1980. FAO Fish. Tech. Pap., (215): 125 p

Fei X (1939) Peasant life in China. E. P. Dutton Company, New York

Friedmann J, Douglass M (1978) Agropolitan development: towards a new strategy for regional planning in Asia. In Lo F-C, Salih K (eds) Growth pole strategy and regional development policy. Pergamon Press. Proceedings of the seminar on industrialization strategies and growth pole approach to regional

Gaubatz P (1995) Urban transformation in post-Mao China: impacts of the reform era on China's urban form. In: Davis D et al (eds) Urban spaces in contemporary China. Cambridge University Press, Cambridge, pp 28–60

Gaubatz P (2008) New public space in urban China. China Perspect 4:72–83

Ju D (2016) 简论农村学校的布局调整. Theory Res 2016(1):182–183

Kropotkin P (1901) Fields, factories and workshops: or industry combined with agriculture and brain work with manual work. G. P. Putnam's Sons, New York

Levi G (1988) inheriting power: the story of an exorcist (Cochrane Lydia G, trans). University of Chicago Press, Chicago

McGee TG (1991) The emergence of *desakota* regions in Asia: expanding a hypothesis. In: Ginsburg N, Koppel B, McGee TG (eds) The extended metropolis: settlement transition in Asia. University of Hawaii Press, Honolulu

Okhitovich M (1930) The notes on settlement theories (Заметки по теории расселения). The Contemporary Architecture (Современная Архитектура) 1930(1–2):7–15

Shan L, Wang C (2015) 撤点并校 的政策逻辑. Zhejiang Soc Sci 2015(3):84–96

Secchi B (1991) La periferia, in: "Casabella", n. 583, Ottobre 1991

Secchi B, Viganò P (2009) Antwerp. Territory of a new modernity. SUN, Amsterdam

Yan R, Gao J, Huang Q, Zhao J, Dong C, Chen X, Zhang Z, Huang J (2015) The assessment of aquatic ecosystem services for polder in Taihu Basin, China. Acta Ecol Sin 35(15):5198–5206

Zhong G, Cai G (1987) 我国基(田)塘系统生态经济模式以珠江三角洲和长江三角洲为例. 生态经济 1987(3):15–20

Zou Y (2013) 略谈江南水乡桥梁的社会功能. In: Zou Y (ed) 明清以来长江三角洲地区城镇地理与环境研究. The Commercial Press, Beijing, pp 187–208

Chapter 4
Layers: Reading the Territory

Abstract To tackle the complexity of the territory, I try to connect the local scale in which the elements are found and the territorial scale, in which the repetition of the elements could create a critical mass. The author created a territorial frame of 50 × 50 km according to a series of conditions for a typical third space, presented by a collection of maps. Several layers are presented, and the selection of the elements for the layers is based on the elements from the Tangqi case study described in Chapter 2. Due to a great lack of data, the author put great effort into collecting historical maps and GIS data and produced a detailed tracing of elements on a territorial scale based on Google Earth satellite images. This means that the mapping of some of the layers—for instance, the trees, the industrial production space, and the Socialist New Villages—has never been carried out before and is truly original, especially on such a large scale. The final part of this chapter includes a reading of the territorial structure via the layers.

Keywords Territorial scale · Repetition · Critical mass · Mapping · Territorial structure

4.1 From Elements to Layers

In the following chapter, several layers are presented via the territorial mapping of certain elements. The repetition of the elements can create a critical mass for tackling the complexity of the territory. This thesis uses the practice of layers by Viganò in "La città elementare" as the main reference. Layers are presented by mapping one or a group of elements, to "identify urban ensembles that remain sufficiently general and can be examined from different viewpoints. They can be related to one another in various ways, bringing to light relationships between them that are not initially evident" (Viganò 1999).

Each layer is themed by one element and explained via several types of mapping: the historical map, mapping of the element on the territorial scale (50 × 50 km) overlapped with a selection of other elements. Taking the element of water as an example: a map of water in the 1930s reveals the historical water network. This is

© The Author(s), under exclusive license to Springer Nature Switzerland AG 2023
Q. Zhang, *The Elemental Metropolis*, The Urban Book Series,
https://doi.org/10.1007/978-3-031-36409-9_4

followed by detailed mapping of the watercourse crossings with the names of the villages, to demonstrate the universal use of a certain water management method. A mapping of the modern fishpond as a specific type of water body reveals considerable territorial transformation. The mapping of water as a fundamental layer is also used in the other layers of elements such as a forest, industrial production, etc. Besides collecting historical maps and GIS data, great effort was put into the detailed tracing of elements on a territorial scale based on Google Earth satellite images, due to the lack of data. Therefore, the mapping of some of the layers—for instance, the trees, the industrial production space, and the New Villages—is unprecedented and original, especially on such a large scale.

The selection of the elements for the layers is based on the elements from the case study of Tangqi described in Chap. 2: water, the forest (orchards), the road, urbanization (peasants' houses), industries, and schools. The intention is not only to use the local knowledge of the elements, but also to illustrate relevantly, but not always conventional, territorial conditions. The discovery of these conditions triggers a completely different understanding of the territory and reveals problems and opportunities on a scale beyond urban–rural dualism. For instance, the "forest" layer reveals an enormous number of trees dispersed all over the territory, both in urban and rural areas, and creates an illusion of a green territory beyond the traditional perception. Our depiction of the layer tries also to go a step further and explore the territorial and structural role of those elements, that are fundamental in the construction and functioning of the space on a local scale. For example, the roads at the county level, i.e., the asphalt roads in the villages, form an evenly distributed network, whose grid has a critical dimension, serving the dispersed small- and medium-sized industrial platforms. When the territorial and structural roles are clarified, those elements become truly, in the term of Gregotti, "morphogenetic elements", that can transform the territory when they are changed. This mechanism is the first step toward a territorial project.

4.2   The 50 × 50 km Frame

The 50 × 50 km^2, bordered by the cities of Huzhou, Huzhou, Tongxiang, covers a representative part of the Yangtze River Delta on many levels (Fig. 4.1). Geographically, it is the main part of the Hangzhou–Huzhou–Jiaxing Plain, the south side of the Tai Lake Basin. Like the rest of the Delta, the plain is low and flat and lies for the most part only 2–3 m above sea level. A huge number of rivers and canals run through it. One of the specificities of the Delta is that the soil holds a great amount of water. The Delta is famous also for its landscape of paddy fields, and the Hangzhou–Huzhou–Jiaxing Plain is entirely organized by the element of *yu*, the typical way of constructing and managing the paddy fields (Fig. 4.2). Besides the Hangzhou–Huzhou–Jangxing Plain, the *yu* areas cover a sizeable area of the Delta, especially between Shanghai and Suzhou, Changzhou and Wuxi, and on the western edge of Changzhou.

|0 | | | | 250km|

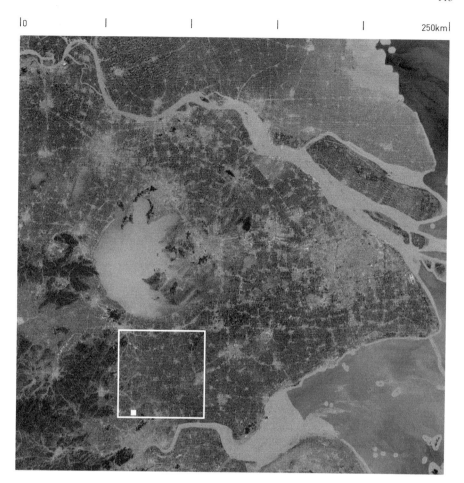

Fig. 4.1 50 × 50 km frame in the context of the Yangtze River Delta and the case study area of Tangqi, 2014. Image: 250 × 250 km; white frame: 50 × 50 km; white dot: 3.2 × 3.2 km, the case study area of Tangqi. *Source* Google Earth. Elaborated by author

At the same time, the 50 × 50 km^2 also holds a slice of the territory that is representative of the third space, in the sense that many of its aspects are in between the urban and the rural. In terms of urbanization, its built-up area is dense and diffuse but without the major city cores or the remote and solitary rural villages (Fig. 4.3). In terms of mobility infrastructure, the whole area is served by a diffuse, intensive, and highly interconnected road system composed of different levels of local roads (Fig. 4.4). There are very few highways and railway tracks going through it. Industrial production is highly mechanized, with an incredible number of village

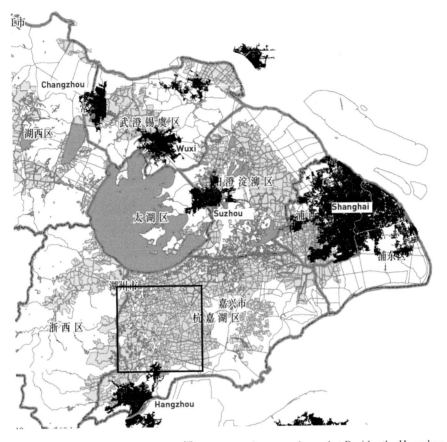

Fig. 4.2 Atlas of the third space: *yu* (圩). The map of *yu* zone shows that Besides the Hangzhou–Huzhou Jangxing Plain, the yu areas spread widely in the delta, especially between the large urban cores: between Shanghai and Suzhou, between Changzhou and Wuxi, and on the western edge of Changzhou. Light gray: the yu zone. *Source* The assessment of aquatic ecosystem services for polder in Taihu Basin (Yan et al. 2015: 5199). Black: the urban cores. Source: GIS data, 2010, National earth system science data sharing infrastructure China. Elaborated by author

industries and small-medium size industrial platforms, but without any national-level special development zones and very few special provincial-level development zones (Fig. 4.5). Concerning energy, there are no ports or local power plants for the import or production of gas, coal, and oil. Contrary to the remote mountain or coastal villages, there are no hydraulic or wind power sources.

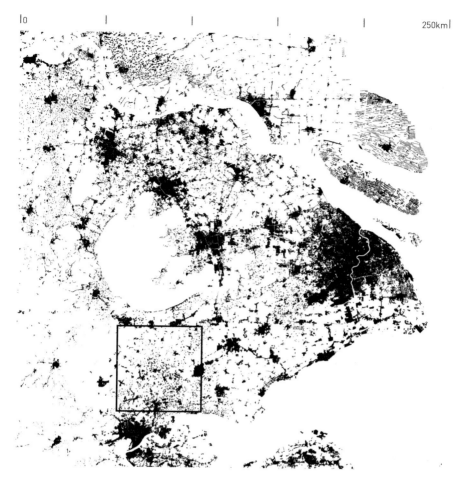

Fig. 4.3 Atlas of the third space: built area. The map of the built-up area shows the dense and at the same time extremely dispersed urbanization, which is different from the urban cores and the remote villages, within the 50 × 50 km frame

The 50 × 50 km scale is part of the protocol implemented within the international research program of the Horizontal Metropolis led by Prof. Viganò at EPFL Lausanne, in which various territories of dispersed urban fabric are studied. The presentation of the 50 × 50 km layer enables a comparison with other cases and ongoing and future work in line with a broader research program.

Fig. 4.4 Atlas of the third space: infrastructure. The map of the mobility infrastructure in the area shows the dense local road networks and the absence of advanced infrastructure (railways and highways) within the 50 × 50 km frame. Red: railways; Black: highways; Gray: other roads. *Source* GIS data, 2010, National Earth System Science Data Sharing Infrastructure China. Elaborated by Author

This 50 × 50 km part of the territory is not the only part of the Delta that has characteristics of "third-ness". On the contrary, it is a segment of a vast and continuous space covering most of the Delta. A series of maps show similar areas within the frame in the southwest of Shanghai, between Shanghai, Suzhou, and Jiaxing, in the north of Suzhou, and between Suzhou, Wuxi, Changzhou, Zhangjiagang, and Changshu.

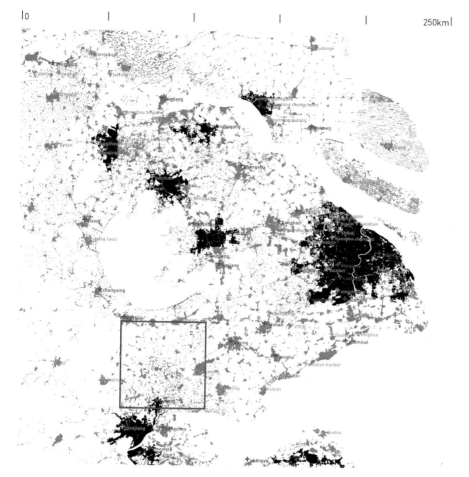

Fig. 4.5 Atlas of the third space: industrial platforms. The map of the industrial platforms shows the lack of large-scale (national and provincial level) industrial platforms within the 50 × 50 km frame. *Source* GIS data, 2010, National Earth System Science Data Sharing Infrastructure China. Elaborated by Author

4.2.1 Layer 01: Water

Water is the most fundamental element in the Delta and its management for agricultural purposes was the main historical motive for constructing the territory. The map from the 1930s (Fig. 4.6) shows a layer of two elements: the water bodies represented by the lines and the towns and villages represented by the dots. The map clearly shows a water network-based diffuse urbanization before the emergence of any major city. The density of the water network grid, with numerous towns and villages at its nodes, is not just defined by the need for water engineering (both drainage and irrigation), but also by the density of human settlements in terms of population, mobility, social activities, etc.

The water network, which runs through or near the cities and the towns, creates physical and socio-economic continuities between urban and rural. Besides the water itself as mobility, logistics, and communication infrastructure, there are also water-related elements—for instance, the bridges. The amazing number of bridges found in both towns and villages is well-recorded in various local historical documents. A

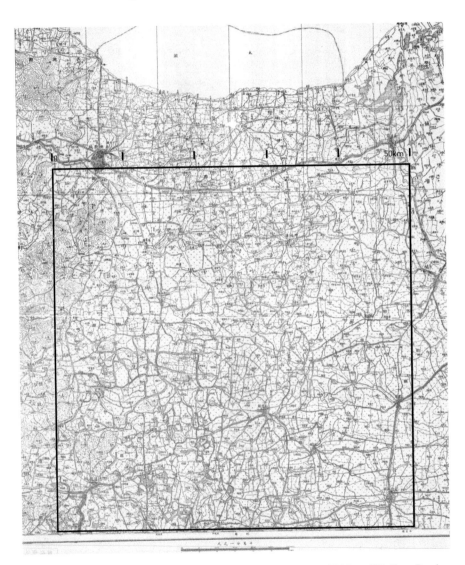

Fig. 4.6 Water layer: the fine historical water network. *Source* 1:100,000 Map of Zhejiang Province (part), Land Survey Bureau, Chinese Army General Staff Office, 1930. Elaborated by the author

bridge creates an interaction between transportation via the water and over land and often fulfills important social functions such as a toll gate, a security checkpoint, and a resting place for travelers. A market or a town often emerges gradually near a bridge. In a typical 1940s town in the Tai Lake Basin, the shops, banks, credit institutions, and services are all located along the water network inside the city and form a cluster around the bridges.[1] Another example is the *yu* system, which determines a local authority. The construction and maintenance of the *yu* require immense local labor and organization, which empowers the numerous local authority and establishes strong local autonomy. The separation dikes of the *yu* were perceived as the division of authority. Therefore, the immense water network also leads to a completely horizontal and isotropic organization of power and social relations, an example of how a political system is strongly connected to the materiality of the territory.[2] Today, this historical horizontality in power and social relations, a significant part of the rurality in the Delta, is gradually losing its material base due to the radical transformation of the territory.

A layer of two existing elements is depicted here (Fig. 4.7): the water bodies, and the villages whose name reflects a traditional water management setup—the construction of *yu*. There are villages all over the territory whose names indicate certain parts of the *yu* (荡, shallow water; 漾/洋, open water; 圩, the plot of land surrounded by dikes; 埭/坝/堰, dikes). Together they represent a historical layer of the configuration of water and a common water management system in the territory. At the same time, the water surface map shows an incredibly diverse landscape created by the implementation of this single technique: canals, especially the small ones, form a dense, evenly distributed grid. In each grid, small waterways divide the land into micro-plots—the *yu*; the fishponds are concentrated on the western part of the map, characterized by egg-shaped ponds, which are usually combined with orchards or other cash crops; the relatively drier and larger *yu* are located on the eastern part of the map and are dedicated more often to rice production.

The *yu* is a stable and flexible structure of land to support a changing production program. Throughout history, the *yu* was used to produce rice, fruit, sugar cane, fish, etc. When the *yu* is used as a fishpond, the wide dikes between the ponds provide space for orchards and other agricultural production—the traditional orchards–dike–fishpond model. The map (Fig. 4.8) shows one of the transformations after the 1980s: a massive change from paddy fields to fishponds. The traditional orchards–dike–fishpond model, in which the proportion of surface between the dike and the pond is about 4:6, is replaced by a modern fishpond-only system with very few dikes, due to

---

[1] See the description of the functionality of the water network in the Ming and Qing Dynasties in multiple studies by Matsuura (Matsuura 1990, 1999, 2001). See the description of bridges, the number of bridges in several counties, and the function of bridges found in multiple official historic books from Zou's study (Zou 2013). A detailed study of the distribution of programs in Linghu Town in the first half of the twentieth century can be found in Yang and Guo's study (Yang and Guo 2013).

[2] See a comprehensive case study of Furong yu in Tai Lake Basin by Sun (2013).

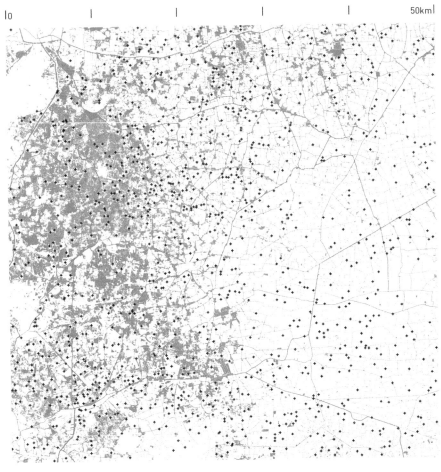

Fig. 4.7 Water layer: villages (red dots) with names related to the *yu* landscape. *Source* GIS data, 2010, National Earth System Science Data Sharing Infrastructure China. Elaborated by Author

the higher economic output of the latter. The modern system, in the form of regular grids in the middle of the *yu*, occupies a large number of paddy fields (Fig. 4.9). The ecological environment is declining: the ponds silt up rapidly because the small dikes are unable to consume the sludge produced by the enormous fishponds and degrade easily because have no natural protection offered by the roots of vegetation on them (Huang 2013).

This transformation caused a territorial struggle: on the one hand, the ecological and cultural value of the orchards–dike–fishpond model has always been well acknowledged; at the same time, the model's abolition seems irreversible due to its significantly lower economic value in comparison to the modern system. On the other hand, its three-dimensional landscape value was well appreciated throughout history

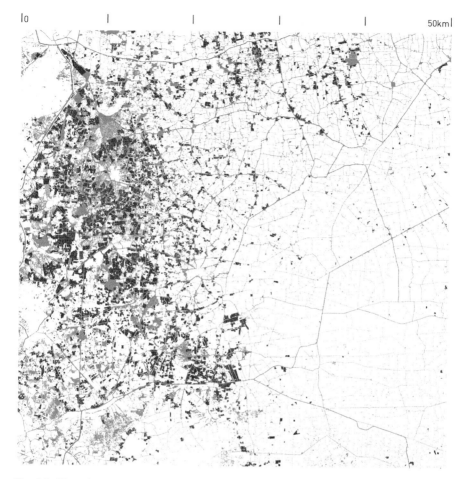

Fig. 4.8 Water layer: the massive introduction of modern fishponds. Black: modern fishponds. *Source*: GIS data, 2010, National Earth System Science Data Sharing Infrastructure China; Tianditu, National Catalogue Service for Geographic Information. Elaborated by Author.

and was a constant theme in poems and prose.[3] Today, the traditional orchards–dike–fishpond landscape model is no longer seen as a livable environment, or, in the current urbanization model, its value has become neglected.

The transformation of the water landscape is certainly not an isolated case but an example of large-scale transformations, which are reported and studied but rarely presented with their territorial impact. At this level, the question is far beyond the

---

[3] Beautiful examples can be found in 宝前两溪志略 "西溪水遛环, 地形若凤, 桑柘遍垄, 桃李成阴, 水际多植 绿杨, 乌桕, 红蓼, 芙蓉, 望如锦绣"; in 彝寿轩诗抄 "村路禽言滑滑泥, 欲行不得怕扶藜. 闲看溪父培桑 本, 泼 泼 光 澈 照 晓 陧"; and in the poem of 释道潜, 送 钱 持 王 主簿 西 归, 松陵接苕雷, 迤逦皆清源". The technique of the orchards-dike-fish-pond model is well documented in 沈氏农书 (The Shen's Book of Agriculture, around 1640).

Fig. 4.9 Zoomed comparison between traditional fishponds (left, 2006) and modern ones (right, 2022) around the Donglin Town within the 50 × 50 study area. *Source*: Google Earth. Elaborated by author

one of a potential loss of the traditional techniques or a poetic landscape scene. It concerns the replacement or the disappearance of the fundamental structural layer of the territory on which the economic, social, and environmental performance of the land was based: the local knowledge, the settlements, the roads, the flood control, the trees, indeed on which the organization of almost all the local elements depended. More radical questions must be asked, the answers to which are extremely open and should not be limited to protectionism: should the traditional fishpond, or even the traditional *yu* landscape be maintained? Who should be maintaining it if the peasants become urbanites? What kind of stability and balance could we design and construct in the territory given the rapid transformation of the last few decades, and considering that the original model was a result of hundreds of years of evolution? Could environmental value even be enhanced by the retreat of or carefully introduced human activities?

4.2.2   Layer 02: Forest

A forest is a large area dominated by trees. According to UN Environmental Indicators, a forest is defined as a land with a tree canopy cover of more than 10% and an area of more than 0.5 ha, including natural forests and forest plantations, but excluding fruit orchards and agroforestry.[4] According to the Chinese standard,

---

[4] *Source* UN Environnemental Indicators. (2010). https://unstats.un.org/unsd/environment/forest area.htm. More detailed specifications are mentioned in the definition of a forest:

"Forest includes natural forests and forest plantations. It is used to refer to land with a tree canopy cover of more than 10 per cent and area of more than 0.5 ha. Forests are determined both by the

a forest is a land with a tree canopy cover of more than 20% and an area of more than 0.667 ha, with the height of the trees equal to or higher than 2 m.[5] According to the official land use data, the Yangtze River Delta has not had many forests. Except for the forests on the few hills, the major part of the land in this area of the Delta is categorized as arable or built land. However, this categorization does not accurately represent the territory, which has a surprisingly large number of trees, covering an area of 44,028 ha,[6] or 17.6% of the 50 × 50 km study area. If all types of trees, including orchards and agroforestry, were to be defined as part of a forest, the entire territory would count as a forest.

The territory is a "light version" of a forest. There are trees everywhere, but their density is relatively low in most parts. The forest does not only consist of trees as a natural element but also includes human settlements, resembling a low-density version of the Boston metropolitan area, where a vast artificial forest was created to accommodate the sprawl of single-family houses. A similar example is a forest in the municipalities of Brasschaat, Schoten, and Schilde on the outskirt of Antwerp, where the majority of single-family-house neighborhoods are located in a forest park with green boulevards and even classic open green axes with fountains—the forest as a framework for living.

A dense layer of the forest, dating from before humans settled in, covered the entire Tai Lake Basin. Orchards have occupied a large part of the territory since the Ming Dynasty (1368–1644) when the production of cash crops started to overtake the production of grain. Orchards, which are categorized as a type of agriculture, have a very unique spatial character, almost like a forest, and contrast with the more traditional agriculture such as rice or cotton fields. The poets of the sixteenth and seventeenth centuries constantly depicted beautiful orchard landscapes.[7] On the map from the 1950s (Fig. 4.10), orchards, which formed a continuous territorial figure with several large interconnected patches, cover a large portion of the 50 × 50 km frame.

Today, this vast figure of historical orchards has disappeared in many parts of the territory due to the transformation of the production method (e.g., the introduction of modern fishponds). However, the extremely diffuse and fine urban tissue has created a diffuse gas-like "forest", that is extremely porous and at the same time extremely

---

presence of trees and the absence of other predominant land uses. The trees should be able to reach a minimum height of 5 m. Young stands that have not yet but are expected to reach a crown density of 10 percent and tree height of 5 m are included under forest, as are temporarily unstocked areas. The term includes forests used for purposes of production, protection, multiple-use, or conservation (i.e., forest in national parks, nature reserves and other protected areas), as well as forests stands on agricultural lands (e.g., windbreaks and shelterbelts of trees with a width of more than 20 m), and rubber wood plantations and cork oak stands. The term specifically excludes stands of trees established primarily for agricultural production, for example fruit tree plantations. It also excludes trees planted in agroforestry systems".

[5] *Source* State Forestry Administration of the People's Republic of China.

[6] *Source* Google Earth satellite image, elaborated by the author.

[7] For example, 沈谦(1620–1670) writes: "水南水北起霜风，蜜橘村村似火红，选得金钱亲剪赠，连枝一对正当中". The types of orchards as the main objects for the poems were the loquat, saccharum, plums, and oranges.

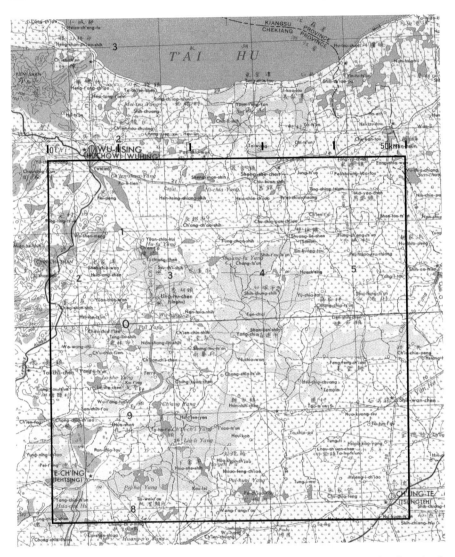

Fig. 4.10 Forest layer: the historical distribution of the orchard's fields. Light green hatch: orchards. *Source* 1:250,000 Map of China (part Hang-Chou), Army Map Service (SNAM), Corps of Engineers, U.S. Army, Washington. D.C. 1953. Courtesy of the University of Texas Libraries, The University of Texas at Austin. Elaborated by the author

light, including orchards, woods, nurseries, street trees, etc. This porous forest, in its different forms, penetrates every part of the urbanized area.

Besides a few trees on hills on the western border of the frame, the forest is artificial: the main part of it consists of orchards and nurseries. Large patches (1000–7000 ha) can be found in the southwest around the Tangqi area, where the fruit orchards predominate; the vast nursery field around the town of Taoyuan in the northeast, and the large nursery field around Yanguan Town and along the Qiantang River in the southeast. The second level of concentration is found around the main cities and

|0 | | | | 50km|

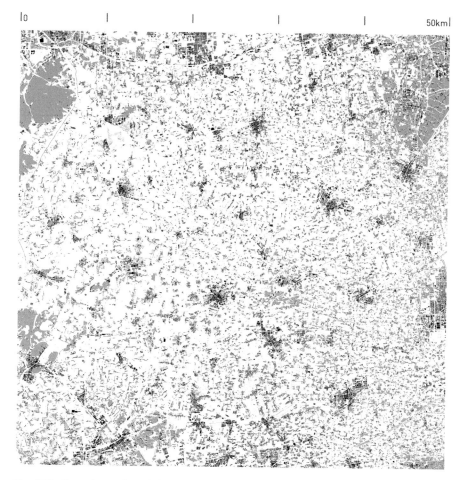

Fig. 4.11 Forest layer: the existing spread and fine layer of trees and orchards around the diffuse settlements. Green: trees (*Source* Google Earth, traced by author); Black: buildings (*Source* Tianditu, National Catalogue Service for Geographic Information. Elaborated by Author

towns—the centers of trade and consumption. Around the city of Tongxiang, where the new urban development—which calls for huge numbers of decorative trees—is taking place concentrically, the nursery fields figures as a "green belt" around the city. The third level of this forest space is a large diffuse pattern of trees, constituted by the small plots of orchards around towns such as Linghu, Hefu, and Donglin that are the traditional centers of the orchards–dike–fishpond model; small orchards and nurseries in each *yu*; the trees—many of which are historical—in the center of each village along the water between the *yu*; and two types of linear continuity of trees: the trees along each dike of the *yu* and the increasing number of street/road trees along the growing mobility network in the territory—both within the cities and in the rural area.

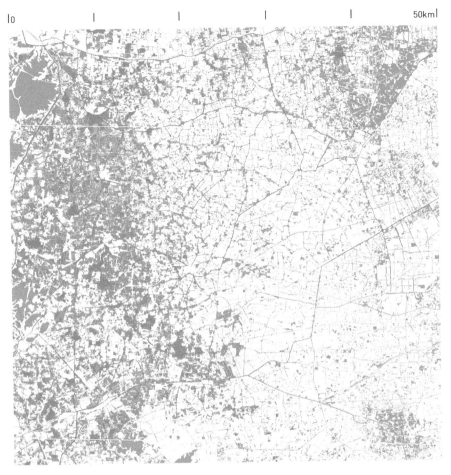

Fig. 4.12 Forest layer: the existing spread and fine layer of trees and orchards overlapped with the water layer. Green: trees (*Source* Google Earth, traced by author); Gray: water (*Source* Tianditu, National Catalogue Service for Geographic Information. Elaborated by Author)

Different levels of tree spaces—comparable to the categorization of the urban green space on the metropolitan, city, neighborhood level, etc. make for an organic system of "forest", with large metropolitan and regional patches, medium plots between the settlements, spaces with small trees within each village, and green corridors connecting the different parts. The distribution of the different levels of space is extremely isotropic and even, and the urban and rural areas share almost the same pattern (Figs. 4.11, 4.13, 4.14, 4.15, and 4.16).

The territory needs reforestation because of environmental deterioration. The forest space is strongly dependent on the water space on the dike, around the ponds, or within the *yu*. The rapid transformation of the territory, the disappearance of villages, and the modernization of the fishponds and their concrete dikes lead to a fragmentation of wooded areas. Nevertheless, the territory still has the potential to regenerate and strengthen local forests into a critical mass.

Fig. 4.13 Zoom of the
forest layer (Fig. 4.11): a
small town is surrounded by
orchards. The white part
within the green area is the
fishponds

The layer of water and trees (Fig. 4.12) indicates a certain direction in this regeneration process: one can imagine a restoration of the orchards–dike–fishpond model whereby the western part of the 50 × 50 frame appears as a large territorial figure of a water–forest space, while the fishponds could be transformed into wetlands, and could become a territorial water purification system.

Fig. 4.14 Zoom of the
forest layer (Fig. 4.11): small
villages are contained in the
large-scale nursery area

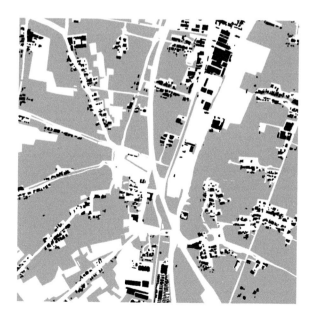

Fig. 4.15 Zoom of the forest layer (Fig. 4.11): the liner tree spaces across both the city and the villages

Fig. 4.16 Zoom of the forest layer (Fig. 4.11): linear tree spaces along the roads and water connect the diffuse villages

4.2.3 Layer 03: Roads

The historical road network has always been diffuse and isotropic. However, it has always been a specific grid that is open, extensive, and dense and serves most of the territory. At the same time, it defines a certain dimension of each grid (approximately 7–8 km), which creates a level of hierarchy. This evenly-dimensioned grid, which consists of roads and tracks, can be seen clearly on the map from the 1950s (Fig. 4.17).

The road system in China has six levels: the highway, the national road, the provincial road, the county road, the town road, and the village road. The highways, the national roads, and railways, of which there are generally very few in this territory,

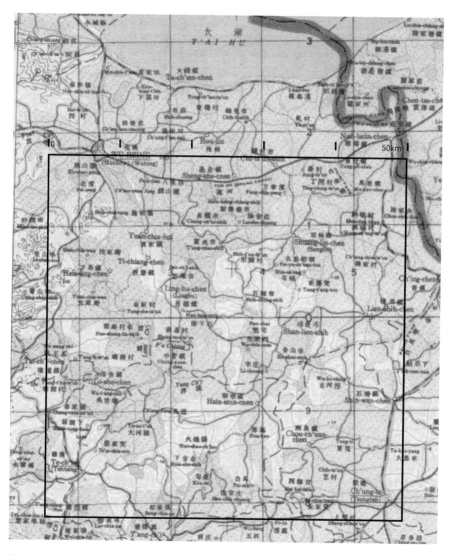

Fig. 4.17 Road layer: the historical layer of roads connecting different cities and towns. Red lines: roads; Red dashed lines: paths. *Source* 1:250,000 Map of China (part Hang-Chou), The Bureau of Measurement, Combined Service Forces, Republic of China 1954

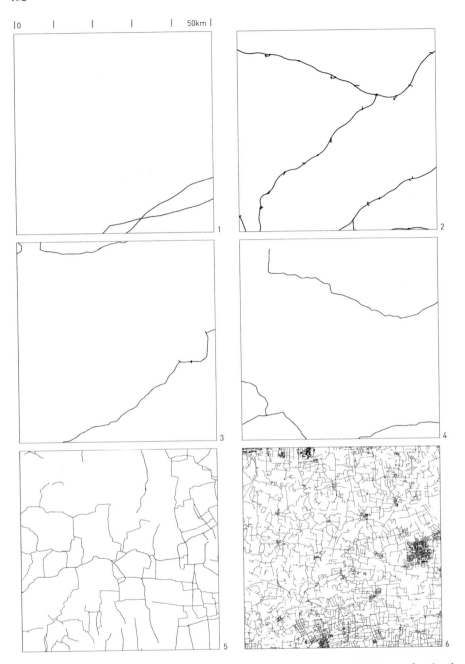

Fig. 4.18 Railway and six levels of roads in China's road system. 1 railway, 2 highways, 3 national roads, 4 provincial roads, 5 county roads, 6 town roads and village roads. *Source* GIS data, 2010, National Earth System Science Data Sharing Infrastructure China. Elaborated by Author

function as "tubes" that run through the middle of the field and are disconnected from the majority of the diffuse urbanity. The town and village-level roads, a very dense grid, reach every corner of the territory, providing basic access to the inhabitants. In the middle of those two layers, the county roads (together with a few provincial roads) play a structural role in the territory. The grid of the county roads, approximately 7–8 km for each grid, is very similar to and even partly overlaps with the road grid of the 1950s, which reveals a continuity of this structure (Fig. 4.18). Today, county roads are the main roads connecting facilities, such as educational institutions, in the territory. Towns, large industrial platforms, and clusters of facilities are located at

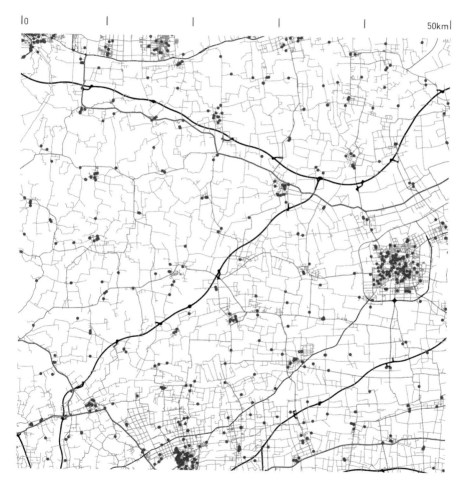

Fig. 4.19 Road layer: the layer of existing roads overlapping with educational institutions. The distance between the provincial and county roads (red) is similar to the one of the historical road networks in the 50 s, which shows a structure with a "stable dimension". Red lines: provincial and county roads; Red dots: educational institutions. *Source* GIS data, 2010, National Earth System Science Data Sharing Infrastructure China. Elaborated by Author

the nodes of this network and show some degree of concentration in this horizontally organized territory (Fig. 4.19).

The stable dimension of this structure could have multiple meanings. It is coherent with the organization of the human settlements throughout history; it is similar to the dimensions of a commune in the Maoist era and to the *agropolitan* district proposed by Friedmann, in which self-sufficient food production is feasible theoretically; it has a radius of 3–4 km or approximately 5 min by bus/tram and 20 min by bike—a comfortable distance for commuting with sustainable mobility; at the same time, it is large enough to have an open landscape between the nodes.

The physicality of the county roads is also crucial. Although in general county roads have different sections, they are not wide according to current standards. The whole section is more than 12 m, and the section for traffic is about 9 m. When they run through an urbanized area, they are easy to change and merge into the local road system. Activities, industrial production, logistics, commerce, restaurants…a variety of activities are now located along the roads. Since the town roads and village roads are often only 5 m wide, and there are few highways and national roads, the county road network is the most suitable for public transport and/or separated bike lanes.

4.2.4   Layer 04: Housing Development: The "New Villages"

A comprehensive overall reading of the rapid and vast urbanization in the Delta is difficult to imagine. Various methods of urbanization are being invented and practiced in the territory: the "city-making" close to the urban cores, the "groups" that are similar to the concept of satellite cities within a critical distance of the main urban cores, the "beautiful villages" in remote rural areas, etc. The map above presents a layer of a particular, but representative element, the construction of "New Socialist Villages" from 2005—a concentrated construction of peasant houses (Figs. 4.20 and 4.21). Its particularity is displayed by its distinctive physical appearance, and its representativeness refers to the typical "concentration first" ideology behind it.

The term "New Village" was first introduced to China at the beginning of the twentieth century and originated in a rural movement in Japan named "New Village". This movement was led by the Shirakaba-ha group, whose members are influenced by Kropotkin's mutual-aid idea. It had a strong influence on Mao when he was imagining self-sufficient rural communities. After 1949, inspired by the Soviet "Kolkhoz" collective farm practice, the "New Villages" were constructed as collective residential areas for workers. Afterward, the term "New Village" was used more generally for the new residential areas, and even today many urban residential areas are called "New Villages". In the 1960s, the term "Socialist New Villages" was introduced, which has a different meaning. The construction of the "Socialist New Villages" was promoted by Zhao Ziyang[8] in Guangdong Province, to modernize and electrify

---

[8] Zhao Ziyang was a high-ranking statesman and the third Premier of the People's Republic of China from 1980 to 1987. He was also Vice Chairman of the Chinese Communist Party from 1981

|0 | | | | 50km|

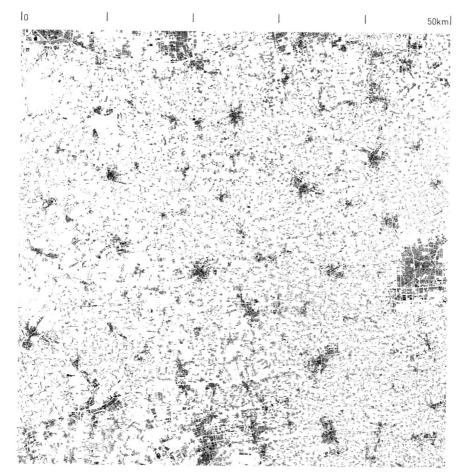

Fig. 4.20 Layer of buildings: the existing "carpet" of buildings in the 50 × 50 km frame. *Source* Tianditu, National Catalogue Service for Geographic Information. Elaborated by author

the rural area. The Cultural Revolution put a stop to this program which was not implemented.

The "Socialist New Villages" today, as a concept, are a further development of the 1960s model. In 2005, the Party officially proposed the construction of "Socialist New Villages" in "The CPC Central Committee's Suggestions formulated by the 11th Five-Year Plan, and part of the coordinated development of the urban and rural areas". It has the following characteristics: it actively promotes the coordinated development of urban and rural areas; modernizes agricultural production; strengthens the general

---

to 1982, and General Secretary of the Communist Party from 1987 to 1989. During his career as the leader of Guangdong Province in the 60 s, he led a series of reforms to save the declining economy due to the predominance of the collective economy. The "Socialist New Village" was introduced against this background.

Fig. 4.21 Layer of buildings: the construction of "Socialist New Villages" in the 50 × 50 km frame. Red: the construction of "Socialist New Villages" since 2005. *Source* Google Earth, traced by author. Gray: all buildings. *Source* Tianditu, National Catalogue Service for Geographic Information. Elaborated by author

rural reform (mostly in terms of economic reform); develops rural facilities (in terms of education, medicine, and basic infrastructure including mobility, communication, and energy systems); and increases the income of the rural population.[9]

Contrary to the all-around and multi-disciplinary content of the term, the physical result is monotonous. It is typically a concentrated construction of a new rural residential district whose inhabitants are the relocated and concentrated rural population from several natural villages. At the same time, the original dispersed village settlements are demolished (Fig. 4.23). In principle, the land that is left after demolition

---

[9] Those aspects were detailed in "Opinions of the CPC Central Committee and State Council on Promoting the Construction of Socialist New Villages" in 2005.

should be reclaimed as arable land, but in reality, it is taken over by the new concentrated construction and its quality and productivity are often more valuable than the compensation received. The design and planning of the New Socialist Villages present an incredibly homogeneous typology[10]: a dense grid of roads with multistory peasant houses, very similar to the ones built spontaneously by the peasants, one next to the other. Within each construction, the typologies of the peasant houses are identical: 3–4 stories high (the maximum volume allowed), usually with a pitched roof (Fig. 4.22). Some of the New Villages are even gated.[11] One could criticize the architectural or spatial quality, but at the same time see it as a much more urban construction: in the cities, these residential districts are standard, with the peasant houses replaced by high-rise buildings. The local concentration of inhabitants also demands new services and facilities, such as small schools that are sometimes built around it. It could be read as a process of micro-centralities starting to merge at the very local scale, in the extremely diffuse and fine layer of settlement throughout the territory.[12]

Three benefits lie based on this concentration: more efficient facilities and infrastructure, the creation of a tidy and orderly community, and the freeing up of constructible land by concentrating the housing plots.[13] Surprisingly, this radical "tabula rasa" concentration operation represents an extremely dispersed development on the territorial scale—a distributed concentration. The scale of each New Village is small, its average size is less than 5 ha, with only about 5% larger than 20 ha. The latter are mostly located around the city of Tongxiang, the rest around the existing towns, next to the villages, along the new infrastructure, in the forest: but in general, evenly across the territory. This dispersed concentration is reminiscent of the idea of "concentrated decentralization" by Armin Meili in the regional plan for Switzerland (Meili 1941). While Meili proposes a network of housing developments of 10,000 inhabitants at the most to form a linear metropolitan area of six million people between Geneva and St. Gallen, the "Socialist New Villages" are concentrations on a similar or smaller scale, but much more local and without emphasis on public transport or the connection between these concentrations. While in Meili's scheme the settlements are connected by public transportation and can benefit from

---

[10] Besides the construction of New Villages, there are other forms of relocating peasants, for example when a peasant house is demolished close to the city, the owner might receive one or multiple apartments in the residential towers as part of the city expansion. We opted for the analysis of the New Village as a specific way of building the territory.

[11] A few exceptions of high-quality architecture in New Villages have emerged, for example the Dongziguan Affordable Housing for Relocated Farmers designed by GAD. Nevertheless, most of the construction remains a massive duplication of the self-build peasants' house.

[12] See also Walter Christaller's Central Place Theory (Christaller, 1933).

[13] The buildable land in rural areas is owned and managed collectively by the rural local authority, and each peasant family is allocated a plot of 100–125 m^2 (in Zhejiang Province) for building its own house, thus the concentration of the peasants' houses, in which each house consumes less land than the individual construction, could save land for economic development of the village itself and even the expansion of the city.

Fig. 4.22 New Village construction as concentrated dense residential development. Shimen town, Yangtze River Delta 2019. Photo by author

Fig. 4.23 Realization (in progress) of the Nanhu New Country Village Master Plan designed by SOM. The sites of the demolished houses were soon recuperated by vegetation, with traces of the foundation in between. Left: original villages and agricultural structure; right: construction in progress. *Source* Google Earth, elaborated by author

city life, the construction of the "Socialist New Villages" is to an extent a reinforcement of the historical dispersion of the population living in rural areas, reinforcement of extreme in-situ urbanization. On the one hand, the historical environments of the villages are being demolished, and the traditional structure of rural society is losing its physical base; on the other hand, it presents an often overlooked ambiguity, that of the Delta moving toward a more concentrated condition as intended by the policies, or actually toward a more robust dispersion, or both, through a series of operations: the New Socialist Villages, the merging of the villages,[14] the urban "groups" (组团, similar to the concept of satellite cities), and even some "city-making".

The construction of the New Socialist Villages demonstrates not only local concentration in a very diffuse way, but also a persistently strong desire to live in the countryside, and an existing and even growing capacity of the environment to support a certain degree of densification via its infrastructure and facilities. The New Socialist Villages are stabilizing the population within and around them in the rural area along with the new roads and telecom towers, and the new quota of the land saved and converted into new activities and industries.

4.2.5 Layer 05: Industrial Production

The Yangtze River Delta has 26 National Economic and Technological Development Zones (2010) and hundreds of provincial ones. Most of those high-level special zones are in cities, especially Shanghai. This model is accompanied by gigantic modern industrial estates, foreign investment, IT or other high-tech industries, dense residential areas with tower blocks, a great number of new migrants, highly developed infrastructure, etc. The most famous example is Kunshan.

This movement has led to concentration. From 2009 to 2014, employment in large enterprises (2000 + employees) grew by 92% in Shanghai, 79% in Jiangsu, and 71% in Zhejiang; in contrast, employment in small enterprises (<300 employees) dropped by 36%, 13%, and 19% respectively. But still, the scale of small industries is immense: the employees in small industries constitute 36% of the total employees in Shanghai, 39% in Jiangsu, and 51% in Zhejiang. In terms of efficiency, the large enterprises are not doing much better than the smaller companies and considering the general industrial crisis in the Yangtze River Delta: the profit margin of the large enterprises is 6–9%, and that of the medium enterprises is 6–8%, and that of the small enterprises is around 5–6%. In 2013, the Kunshan development zone reported a profit margin of 5%, with the IT and high-tech industry accounting for most of this. In some industries, this tendency toward concentration has even reversed. The

---

[14] Since the announcement of "Decision on deepening reform and strict land management" by the State Council in 2004, the merging, demolition, and renovation of villages have been taking place all over China. See the study on the transformation of the administration in this process (Wu 2014).

number of textile factories in Jiangsu increased from 5555 to 8239 from 1989 to 2009, in Zhejiang from 5986 to 8743 while the average number of employees dropped from 303 to 143 in Jangsu, and from 164 to 134 in Zhejiang.

This very strong layer of small industries echoes what Kropotkin found in England at the end of the nineteenth century—a diffusion that opposes concentration or even cooperates with concentration. McGee described this new tendency:

> These circumstances have led more recently to an emerging extended metropolitan region that is beginning to develop industrial clusters and functional specialization between urban centers within the lower Yangtze River Delta. At the same time, rural town and village level industrial production continues to be important, especially as it becomes increasingly linked to larger scale local enterprises (McGee 2007: 162).

Spatially, there are vast areas in the third space without national or provincial industrial zones. There, as the mapping of industrial space in the 50×50 km frame shows (Fig. 4.24), each town has its small industrial estate; small industrial platforms are located along the main road (county road for example); factories can be found spread over the fields, and family workshops on the ground floor of peasant houses or in their backyard. In the study area, these diffuse industries occupied 24% more space than they did in 2009. Their growth is on all scales all over the territory.

Within this all-scale, diffuse distribution of industry, two categories can be discerned: one of the micro-scale industries irrigated by the capillaries of village roads, and one of the "nodes" or small industrial platforms backboned by the county road-scale mobility network and the main waterways. This combination of decentralization and a certain level of concentration is comparable to the distribution of industry in Flanders. Only in the case of the Yangtze Delta, the amount of industrial space is larger but even more decentralized (Figs. 4.25 and 4.26).

If the micro-industries and the village roads can be read as a fine carpet of production (Fig. 4.27), the small industrial platforms can be read as nodes—not just as economical nodes for trading, the introduction of new technology, connections with global markets and raw materials, and the distribution of energy, but also as spatial nodes, which have comfortable service ranges (the same as the county road, or smaller) to rapidly communicate with the micro-industries and deal with output and orders, with a relatively short daily commuting distance for the local peasants (Fig. 4.28). This cooperation becomes even closer when the employee of the micro-industrial platform lives in a village like the business he works for. The industrial nodes, which often are at the same time the location of markets, schools, and other activities and facilities, embody a strong continuity along the county road—the nodes, the linear road trees, the public transport (buses), the activities and facilities, together compose a "boulevard" in the third space, a civic axis.

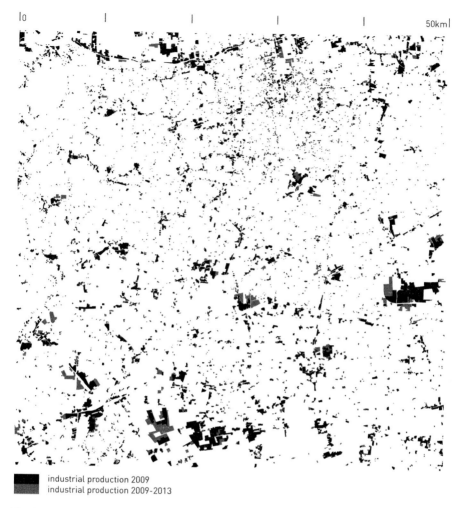

industrial production 2009
industrial production 2009-2013

Fig. 4.24 Layer of industrial production: the existing and growing layer of dispersed industrial production space. *Source* Google Earth, traced and elaborated by author

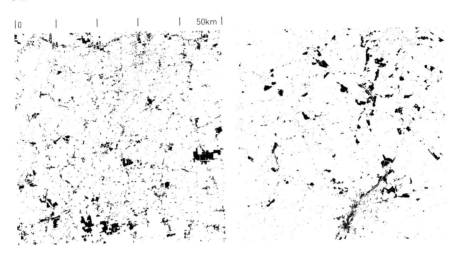

Fig. 4.25 Comparison of industrial production in the 50 × 50 km "third space" and the one in Flanders. Left: the industrial space in YRD, 19,463 ha. Right: the industrial space in Flanders, 13,354 ha. *Source* GIS data, 2010, National Earth System Science Data Sharing Infrastructure China; and CORINE Land Cover 2000, European Environment Agency. Elaborated by Author

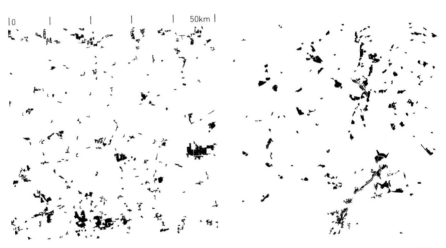

Fig. 4.26 Comparison of industrial production "nodes" (larger than 10 ha) in the 50 × 50 km "third space" and the one in Flanders. Left: the Yangtze River Delta, 12,883 ha, 66% of the total industrial area; Right: industrial spaces 10 + ha, Flanders, 10,367 ha, 78% of the total industrial area. *Source* GIS data, 2010, National Earth System Science Data Sharing Infrastructure China; and CORINE Land Cover 2000, European Environment Agency. Elaborated by Author

|0 | | | | 50km|

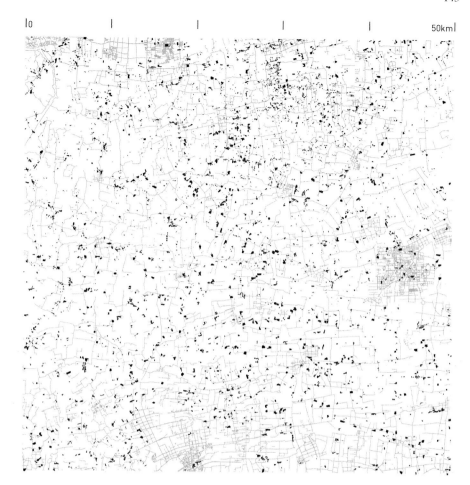

Fig. 4.27 "Carpet" of micro-industries in relation to the diffuse and fine network of town and village roads. Black: micro-industries. *Source* Google Earth, traced and elaborated by author. Gray: town and village roads. Source GIS data, 2010, National Earth System Science Data Sharing Infrastructure China. Elaborated by Author

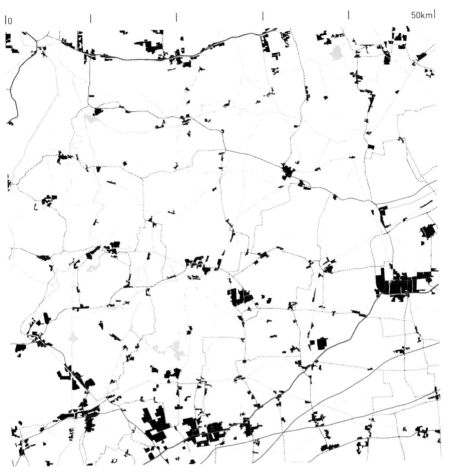

Fig. 4.28 "Nodes" of small industrial platforms in relation to the country roads and main canals. Black: small industrial platforms. *Source* Google Earth, traced and elaborated by author. Red: county roads, source: GIS data, 2010, National Earth System Science Data Sharing Infrastructure China. Elaborated by Author. Gray: main canals, source: Tianditu, National Catalogue Service for Geographic Information. Elaborated by author

4.2.6 Layer 06: The (Elementary) School

More efficient utilization of services, for instance, schools, has been used as one of the main arguments for concentration. Beyond that, the type, scale, and distribution of the facilities such as schools and a model or a plan of urbanization can mutually determine each other. This relation between services (schools) and urbanization is strongly presented in the Central Place Theory by Christaller, the *Broadacre City* project by Wright, *The Ideal Communist City* by Gutnov, the *Agropolitan development and the Modular city* by Friedmann, the *Vision for Brussels 2040* by Secchi and Viganò, and

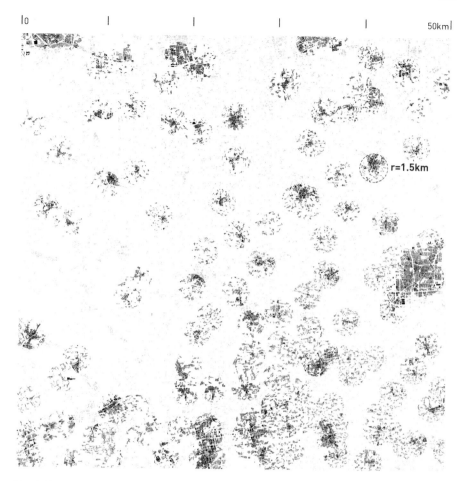

Fig. 4.29 Layer of elementary schools: The walkable distance (1.5 km) from the elementary schools to it surrounding urban tissues. Red: elementary school. Black: buildings within 1.5 km of the elementary schools. Gray: the other buildings. *Source* GIS data, 2010, National Earth System Science Data Sharing Infrastructure China; Tianditu, National Catalogue Service for Geographic Information. Elaborated by author

many others. This relation has also been increasingly noticed and studied by Chinese scholars especially after the "decommission[ing] of the educational stations and the merg[ing] of the schools" (撤点并校) that began at the beginning of the twenty-first century.[15]

The elementary school system in the Yangtze River Delta is decentralized, because of the radical intervention of building a school for each village during the Maoist era.

---

[15] See the study on the relation between the distribution of the schools and the rural settlements (Zhou 2016); see the study on the relation between school and urbanization under the recent new policy of "new urbanization" (新型城镇化) and "two-child" (二孩) (Liu 2015).

In the twenty-first century, the number of elementary schools has reduced rapidly due to the concentration of the schools. For example, in Zhejiang Province, the number of elementary schools dropped by more than 40%, from 5471 in 2006 to 3269 in 2016.[16] In the great urbanization wave and demographic redistribution, the necessity and the benefits of concentrating on schools have been broadly recognized (Liu 2015; Zhang 2015; Ju 2016). Nevertheless, concentration has hit an obstacle: the speed of concentration has slowed down (Li 2017; Zhang 2015), and the negative aspects have begun to be recognized and studied (Wang 2017; Li 2015; Guo 2017; Shan and Wang 2015), even the necessity and feasibility of reintroducing a more dispersed system of schools is being examined (Jia 2016). One of the negative aspects most often stated and a result of the concentration of schools is the reduced accessibility, which is difficult, especially for the students at elementary schools. Although the studies are mostly on remote rural areas where the accessibility problem is more evident, in the third space, the possibility for each student to walk to his/her school is gradually disappearing. If a 1.5 km distance from the student's house to his/her school is considered walkable, a significant number of inhabited areas are not served. Many of the villages, whose elementary schools have been merged into bigger schools, are experiencing demographic growth.[17] For instance, the population of Dinghe village in Tangqi increased from 2548 in 2006 to 3090 in 2012, but its two schools (Zhujiajiao School and Dinghe School) were demolished in 2010 and the students now must attend the Third Elementary School of Tangqi, which is 3 km away.

The walking distance analysis reflects a phenomenon in the third space that is very similar to the "New Socialist Village", a "concentrated decentralization" (Fig. 4.29). The concentration has been accomplished to a certain extent, but the global image is still diffuse. The distribution has reached a critical level whereby the evidence of its benefits is reducing, while the resistance from the local community is increasing. In the third space, the scale and distribution of the schools that function as a "central place" in Christaller's term is becoming embedded, along with a population pattern. Planning questions start to surface as the population and the number of school-age children are increasing in rural areas. Should the investment be in fewer but larger schools, or smaller but many more schools? Should the population be further concentrated and densified around the schools, or do the schools have to be more dispersed to improve accessibility? Does another type of mobility (for example, public transport) need to be developed (Fig. 4.30)?

At the same time, the potential quality of the schools in the third space remains hidden. The mapping of green spaces (spaces with trees) within 1.5 km of elementary schools (Fig. 4.31) is just one example. It shows that many schools are located in large-scale orchards or nursery fields, or have green spaces nearby that are much larger than the school itself. The combination of high-quality natural space, schools, and their dormitories is imagined and promoted by utopian thinkers like Howard, Wright,

---

[16] Source: The yearly official report on educational development in Zhejiang Province, from 2007 to 2016.

[17] See the increase in the number of school-age children as a common condition in the northern Jiangsu Province (Wang 2017).

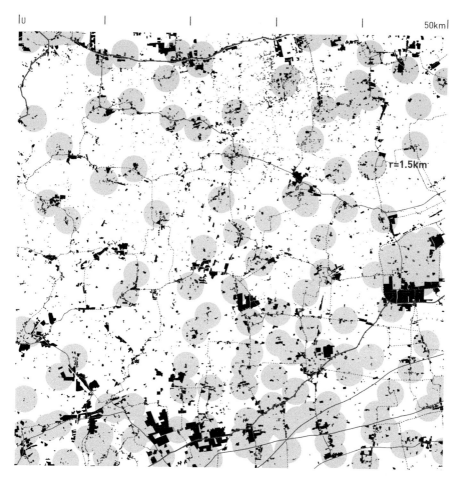

Fig. 4.30 Layer of elementary schools overlaps with industrial production and county-level roads. It shows the potential to combine education with industrial production and the future public transport network. Red: elementary school. Black: Industries. Red dash lines: the county road network. *Source* GIS data, 2010, National Earth System Science Data Sharing Infrastructure China; Tianditu, National Catalogue Service for Geographic Information. Elaborated by author

Okhitovich, and Ginzburg.[18] Design questions could be asked: Can we imagine such a combination in the third space? Can we imagine a different type of education? Or more radically, how should a school-age child live, study, move, and play in such a setting?

---

[18] See a small study on this idea of schools in this chapter.

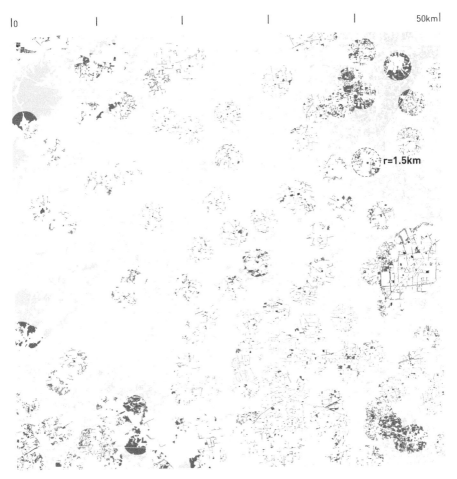

Fig. 4.31 Layer of elementary schools and the forest in the school's walking distance (1.5 km).
Red: elementary school. Drak blue: areas covered with trees within 1.5 km of the elementary schools
Light blue: the other trees. *Source* GIS data, 2010, National Earth System Science Data Sharing
Infrastructure China; Google Earth, elaborated and traced by author

4.3 A Conclusion

The exploration of the layers, almost a strict repetition, and continuation of the
exploration of the local elements at a territorial scale, does not intend to construct a
comprehensive understanding of the territory; rather it is an attempt at exploring the
territory from a certain perspective—its elements. Nevertheless, three conclusions
can be drawn.

The first is on the very material basis of the territory. This very limited exploration
already represents an unconventional view of the reality of the basic construction of
the territory. For example, although the mechanism of the *yu* and the orchards–dike–
fishpond model have been studied for decades, their scale and territorial form have

not been depicted; the density of trees, which addresses this third space of the Delta as a "green territory", has never been explored. At the same time, this limited study also reveals a few of the transformations, hidden but also incredibly massive and rapid,[19] of this material base, such as the disappearance of the traditional fishpond, and the growth of dispersed industries.

Those discoveries show a capacity and necessity for such exercises, which could generate territorial knowledge and even subvert an entirely different perception and image of the territory. Thus, the traditional image of the third space as a rural scenery, a stationary agricultural field, an almost empty palimpsest ready for any new intervention, is radically changed into one of a territory that has many layers, full of content, and rapidly being transformed and modernized.

The second insight is on the issue of concentration versus decentralization, and the rural versus the urban. While the main discourse has been on how to choose between those opposites, this limited exploration has attempted to uncover the extreme complexity of the process that is often about concentration and decentralization, and urbanization and ruralization, one embodied in the other, at the same time. First, many concentration and urbanization processes start at a very local scale, such as the Socialist New Villages and the small-medium industrial platforms, the first degree of centrality over the pixels of houses and factories. When illustrated on a territorial scale, this type of concentration appears incredibly decentralized in comparison with other processes, such as the expansion of the urban core of Shanghai, or the urbanization along the corridor between Wuxi and Suzhou. Second, this limited exercise implies that, in the third space, those concentration and urbanization processes have reached a critical extent, that they have significantly slowed down or even stagnated because of the reduced marginal benefit and resistance from the communities. The exploration also implies a degree of detachment from the new level of concentration and urbanization, which is purely based on rational concepts such as functionality and efficiency, and the historically more dispersed environment, which is constructed according to the specificity of the territory. Such detachment can be found in the growing commuting distance for the students of elementary schools, and the built-up and agricultural land abandoned after the construction of Socialist New Villages, as well as the deconstruction of the original villages; such detachment also demonstrates the gap between, in the idea of Viganò (Viganò 2013), the two "rationales"[20]—"territorial rationale" and "geometric rationale". A further continuation of the layer exercise could prove such a hypothesis: the struggle

---

[19] The history of the stunning scale of modern fishponds is only twenty years old; in ten years, the diffused industrial space will grow by 70% if the current pace is sustained.

[20] The idea of "territorial rationale" and the "geometric rationale" were used by Viganò in the diagram by Erich Gloeden (1923) describing the city of millions of inhabitants as a set of equivalent cells arranged horizontally and crossed by a river. She uses the term "territorial rationale" as the principal elements such as the water and landforms of the given territory were used as defining elements in the diagram, and the term "geometric rationale" as the inhabited cells are organized by the rhombus pattern.

and collaboration between the two rationales provide a certain level of concentration and efficiency, preventing the ever-growing concentration and urbanization, and stabilizing the generality of dispersion and rurality.

The third insight is an interpretation of the territory. Based on the construction of certain layers, this thesis attempts to recognize a territorial structure within the high degree of diffusion. The initial point is not to find unity in the fragmented parts—on the contrary, it is difficult to call the territory fragmented: the layer of each element presents a strong order based on its own history and rationality. It is essential to capture a structure (Fig. 4.32)—or more precisely, a group of certain elements with a particular scale, through which the new transformation and quality of the territory can be imagined. This interpretation is also made by connecting different layers to reveal the relations between different elements, which define the structural and figurative elements. The interpretation is an implicit project in which some aspects of the elements are preserved and given new meanings, and others are discarded, a process through which the link between the existence and perspective is continuously created and updated, a process as a step toward a project.

Network of canals and roads (Fig. 4.33, A).

The network of roads (the county roads) and the main canals create a network penetrating both the urban and rural environments. The treatment of the banks along the water bodies, the trees along the roads, the sidewalks, the bike lanes, the public transport (tram, bus, waterbus), the street furniture (trash cans, bus stops, benches, bike stands), the material of the surface, the language of the architecture along it—a full spectrum of elements related to this network could create a spatial continuity between the urban and the rural.

Nodes of industrial production (Fig. 4.33, B).

The industry is changing. In the Yangtze River Delta today, the traditional heavy and manufacturing industries are feeling pressure to relocate due to the increase in labor and land costs. At the same time, automation, the internet, and artificial intelligence will significantly reduce the space that is needed for production, storage, and logistics. The optimization of the industrial space is leaving behind land (sometimes polluted), abandoned buildings, and infrastructure. The micro-industrial platforms are located at the nodes or along the network of canals and roads, the transformation of which is strategic.

The territorial figure of the fishpond-orchard (Fig. 4.33, C).

A large continuous space combining fishponds and orchards can be seen on the western part of the map, the scale of which can imply particular opportunities. It occupies approximately 1/3 of the frame and a surface of 650 km^2—a scale that is comparable to regional parks around the world such as Parc Naturel Régional du Vexin Français in Paris (710 km^2), Parco Agricolo Sud Milano in Milan (470 km^2), Emscher Landscape Park in the German Ruhr region (800 km^2), and the country

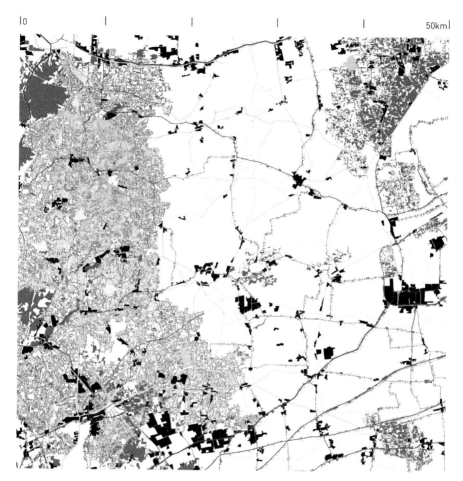

Fig. 4.32 An interpretation of a territorial structure. Black: industrial plots 10 + ha. Red: county roads. Gray: large patches of fishponds and main canals. Green: large patches of orchards/forest. *Source* GIS data, 2010, National Earth System Science Data Sharing Infrastructure China; Tianditu, National Catalogue Service for Geographic Information. Google Earth, elaborated and traced by author

parks in Hong Kong (440 km^2). If it could be entirely transformed into a wetland based on its fishpond structure, it could be compared to the scale of Parc Naturel Régional de Camargue in France (820 km^2)—a wetland of international importance.

Patches of forest (nursery fields) (Fig. 4.33, D).

The patches of forest are located around important cities and towns, spreading on the eastern part of the frame. These patches, although mixed with a certain degree of urbanization, measure around 1000–7000 ha—a dimension similar to a large urban park (e.g., Bois de Vincennes, Paris, 996 ha; Het Amsterdamse Bos, Amsterdam,

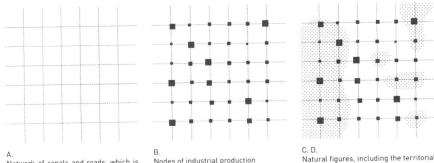

A.
Network of canals and roads, which is supporting a diffuse and isotropic urbanization.

B.
Nodes of industrial production

C. D.
Natural figures, including the territorial figure of the fishpond-orchard and the patches of forest.

Fig. 4.33 Diagram of an interpretation of the structure of the territory

1000 ha), to a regional and territorial green space (e.g., the Zoniënwoud, Belgium, 4421 ha; De Biesbosch National Park, the Netherlands, 9000 ha; Nationaal Park De Hoge Veluwe, the Netherlands, 5500 ha).[21]

References

Christaller W (1933) Die zentralen orte in Süddeutschland. Jena, Gustav Fischer

Guo K (2017) On the reason of implementation deviations of rural school layout adjustment policy and its solution. J Guangdong Univ Educ 37(4):5–12

Huang L (2013) 湖州市桑基鱼塘现状与发展建议. [online] 浙江大学湖州市现代农业产学研联盟. Available at: URL http://www.zjuagri.net/redir.php?catalog_id=422&object_id=11073. Accessed 30 Sep 2017

Jia Y (2016) Analysis on the necessity and feasibility for retrieving rural teaching points in the post school-mapping era—based on the survey in rural areas of nine provinces in China. J Central China Normal Univ (human Soc Sci) 55(1):149–159

Ju D (2016) 简论农村学校的布局调整. Theory Res 1:182–183

Li Y (2015) 农村学校的撤存之争: 农村学校布局调整的方向. 继续教育研究, 2015(10):38–40

Li N (2017) 农村学校布局调整政策之批判分析. 当代教育科学. 2017(2):63–67

Liu S (2015) New urbanization, selective two-child policy and new trends of school layout and adjustment. J Northeast Normal Univ (philos Soc Sci) 276(4):187–191

Matsuura A (1990) 有关明代江南的水运. In: Yamane Y (ed) 山根幸夫教授退休明代史论丛. Tokyo, 汲古書院

Matsuura A (1999) 有关清代州的水运. 关西大学文学论集, 48(3)

Matsuura A (2001) Inland river transportation in Jiangnan in the Qing Dynasty. Stud History, 0(1):35–41

Meili A (1941) Landesplanung in der Schweiz. NZZdruck

Shan L, Wang C (2015) "撤点并校"的政策逻辑. Zhejiang Social Science. 2015(3):84–96

---

[21] Comparison between the nursery fields and parks is not irrelevant. The quality of a nursery field can be appreciated and combined with cultural, sport, and leisure programs. Nursery parks as a particular typology of urban parks can be found in many cities all around the world, for instance, the West Kowloon Nursery Park in Hongkong, Parc de la Pépinière in Nancy, the Pistoia Nursery Park in Italy, and the Nursery Parks in Jinan, Chengdu, and many other Chinese cities.

Sun J (2013) 圩田环境与江南地域社会—以芙蓉圩地区为中心的讨论. In: Zou Y (ed) 明清以来长江三角洲地区城镇地理与环境研究. The Commercial Press, Beijing, pp 402–420

Viganò P (1999) La città elementare. Milano: Skira

Vigano P (2013) Urbanism and ecological rationality. Two Parallel Stories", in Pickett STA, Cadenasso ML, McGrath B (eds) Resilience in ecology and urban design, Springer

Wang A (2017) 农材学校合理布局机制研究—以苏北地区为例. 当代教育科学, 2017(3):19–23

Wu K (2014) The transition of rural space in Zhejiang Province. Develop Reform Res 288(5):1–24

Yan R, Gao J, Huang Q, Zhao J, Dong C, Chen X, Zhang Z, Huang J (2015) The assessment of aquatic ecosystem services for polder in Taihu Basin China. Acta Ecol Sin 35(15):5198–5206

Yang W, Guo T (2013) 明清以来菱湖镇的空间与结构—以产业, 金融, 商业等为中心. In: Zou Y (ed) 明清以来长江三角洲地区城镇地理与环境研究. The Commercial Press, Beijing, pp 77–106

Zhang M (2015) Critique of school merging policy implementation in urbanization of China. Educ Sci 31(3):31–35

Zhou J (2016) 村镇小学布局与居民点体系的互动关系研究. Master thesis. 中国城市规划设计研究院

Zou Y (2013) 略谈江南水乡桥梁的社会功能. In: Zou Y (ed) 明清以来长江三角洲地区城镇地理与环境研究. The Commercial Press, Beijing, pp 187–208

Chapter 5
Systems: Construction of a Utopia

Abstract This chapter has two parts. The first part, "Atlas of a utopia", is dedicated to constructing an atlas of works by a small group of thinkers and designers, including Kropotkin, Frank L. Wright, the Soviet disurbanists, the Japanese Shirakaba-ha group, and Mao, not as the work of a historian, but as an exercise to find a common social and spatial agenda among them toward a unity of the urban and the rural. The atlas is used as a reference utopia for the Yangtze River Delta. The second part, "Imagining a utopia", is an exercise in imagining specific qualities of life in a society where the urban–rural divide is eliminated by transforming the elements, instead of providing concrete solutions. This is accomplished using a set of systems, in which the social and spatial agenda is imagined within the possibilities offered by the specificity of the space of the territory. Part of this chapter has been published as the article "Zhang, Q. (2018). Atlas of a utopia. In: Velo, L., Pace, M. eds., *Utopia and the project for the city and territory*. Rome: Officina Edizioni".

Keywords Utopia · Social and spatial agenda · Imagination · Systems

5.1 From Elements and Layers to Systems

If the objective of describing the elements is to identify the physicality and mechanism of each element as a single entity and if the objective of describing the layers is to discover complex relations through the repetition of a selection of elements on the territory—a critical mass, the systems are the steppingstones toward a project.

The concept of zoning, which is still the main theory behind urban planning in China, has been increasingly criticized. In Europe and the United States, urbanists develop more and more theories and projects through systems instead of through zones. Although the critique of the former can be traced back to the 50 s, it was from the time of the Bergamo Plan (Secchi and Gandolfi 1992–1994)[1] when the shift from zoning to systems started. Since that time, thought processes and projects take

---

[1] The project focuses on systems rather than zoning to structure the plan. Credit: Bernardo Secchi and Vittorio Gandolfi—with P. Cigalotto, M.G. Santoro, and P. Viganò team coordinators; consultants: L. Caravaggi and A. Tomei; P. Gabellini; and D. Rallo. The plan rejects "the great analogy of the

different paths, including Seoul City Airport Master Plan by OMA and Drosscape and Systemic Design© by Alan Berger.[2]

Viganò states, "the structure that enables us to pass from an accretion of objects to a system is not preordained or given but is produced by a mental construct as the outcome of a project that only becomes precise and specific over time" (Viganò 1999). Instead of scrutinizing system theories, two observations mentioned by Viganò (tested by Viganò and Bernardo Secchi in the construction of urban projects and plans) draw particular attention: the system as an ensemble of elements and the system as an ensemble of performances. As she points out, the system is an ensemble of elements, the concept of the system "can make reference to concepts of identity and belonging that is, the nameability and recognisability of each element and its belonging to a group, a whole, a family". The system is also an ensemble of performances: "the point of designing with systems is to define some levels of consistency and integration for all the parts of the city by superimposing multiple functional programs and making them intersect, searching for non-global consistencies that relate to the attainment of an articulated ensemble of performances". Therefore, the design process could start not only through the introduction of new types of elements but also through the redefinition of the identity and the performance of the known elements for the invention of a new system.

It is fascinating to see how in some projects—both social and spatial—a radically new system can be imagined through the redefinition of the identity and/or the performance of certain elements. The entire Soviet settlement—the replacement of the cities—was imagined by the disurbanists through their redefinition of the house itself. Broadacre City by Wright was imagined through the redefinition of a series of elements and their particular performances in the new systems: for example, the performance of the new intersection and the highway overpass is crucial to the mobility system.

Today, new systems must be imagined by confronting the limit of concentration in the urbanization process on a specific type of territory. My imagination is contextualized and inspired by a group of historical visions by a group of designers and thinkers. In those visions, I could find similar challenges which the visions were constructed for, similar types of territory in which the visions were established, similar spatial systems designed for those visions, the similar performance of those systems—or similar social and spatial agendas embodied in those systems. These similarities, on many occasions identicalness, show an inner connection and necessity that certain territories are calling for certain social-political agenda, which can be only realized within certain spaces. Therefore, the form of territory, the challenges, the visions, the systems, the performances, and the social and political agendas, from different authors, can be connected and mapped on an atlas—an atlas of utopia, as those

---

city as a machine, which had inspired the modern 'functional city'" (Viganò 2012: 410). See also *La Macchina Non Banale* (Secchi 1988).

[2] See the chapter "From Zoning to Systems" in Viganò's article "*Urbanism and Ecological Rationality* the path from zoning to systems in urbanism, and the directions of different practices" (Viganò 2012).

visions are often recognized as utopian. From there, my own imagination begins, as an addition or extension of the atlas.

5.2 Atlas of a Utopia

The atlas of a continuous utopian thinking process by Kropotkin, Mao, the disurbanists, Wright, and many others can be used to illustrate a combination of social and spatial theories (Fig. 5.1). Individually, each treatise contributes one or several aspects to the combination, but collectively, the inner coherence of those aspects allows connecting them and building an entirety toward a common direction. It is an atlas of thoughts, which criticizes the imbalance in urban and rural development because of the dependency on industrial export and the neglect of agricultural production. It discovers and reflects on a certain type of territory—a dispersed inhabited agrarian-industrial territory with a dense layer of small industries and workshops in the fields. Confronting challenges such as urban congestion, the migration from the rural to the urban, the hygiene, and health issues (mainly in the cities), the atlas imagines a new socio-political agenda within a well-conceived space and valid for territories like the Yangtze River Delta where similar questions arise and answers are needed urgently.

The inner coherence of the works of people like Kropotkin, Mao, the disurbanists, Wright, etc. stems also from the direct influence they had on each other. The objective

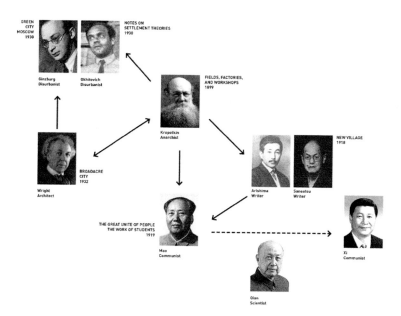

Fig. 5.1 Atlas of "utopian" architects and thinkers. Elaborated by author

of this atlas is to reflect on some of Mao's utopian ideas, especially his vision of the new village, the industrialization of the countryside, and the new education. The atlas aims to amalgamate the thoughts of Mao, Kropotkin, the disurbanists, Wright, Friedmann, Saneatsu, Qian, and others into a network of thoughts.

5.2.1   From Kropotkin to Mao

In his conversation with Edgar Snow in 1936, Mao remarked that he had been strongly influenced by anarchism (Snow 1937: 149). During the May Fourth Movement which took place from 1917 to 1921, he worked as a librarian at Beijing University, where he had the opportunity to read Kropotkin. In his article "The Great Union of the Popular Masses", he favors Kropotkin over Marx and promotes the former's Mutual Aid theory. In 1920, Mao opened a bookshop called "the bookshop of culture", which sold books by Russell, Kropotkin, Darwin, etc. The bookshop became a base for communist party activities, and the "Study of Russia group" was founded there.[3]

It is not that unusual that Mao was interested in Kropotkin. Between 1905 and 1920, anarchist thinking was a vital part of the intellectual protest movement in both Japan and China (Scalapino and Yu 1961). Moreover, the ideas of Kropotkin also influenced Mao through the Shirakaba-ha group in Japan—an influence that served less as a theory than as a practical method. Shirakaba-ha was a literary magazine published from 1910 to 1923. It introduced Western and Russian (especially Tolstoy's anarchist thinking and semi-rural lifestyle) literature and art to Japan. Marx, Engels, Kropotkin, and the Japanese leftist movement had also an influence on the "Study of Russia group". From 1903 to 1907, Arishima Takeo, one of the founders and key figures of the group, "studied in the United States, where he came across the socialist ideas of the Russian anarchist Peter Kropotkin" (Yiu 2008). In the spirit of mutual aid, he created a "Kaributo Cooperative Farming Organization" on his own farm. It is easy to imagine that the ideas of Kropotkin were discussed and disseminated during the Shirakaba-ha meetings with his colleagues. In 1918 when he heard Saneatsu was preparing his Atarashikimura 新しき村, "New Village", Takeo wrote an open letter to express his approval of Saneatsu's scheme. The idea of the "New Village" is community-based and a combination of agriculture and art. In the "New Village", each individual works compulsory six hours a day and can spend the rest of the day freely in "any form of personal interest that leads to the actualization of the authentic self".[4]

---

[3] See a short history of the bookshop from Phoenix Weekly: http://www.ifengweekly.com/detil.php?id=2774.

[4] A similar organization can be found in More's Utopia:

"…but they (the Utopians), dividing the day and night into twenty-four hours, appoint six of these for work, three of which are before dinner and three after; they then sup, and at eight o'clock, counting from noon, go to bed and sleep eight hours: the rest of their time, besides that taken up in work, eating, and sleeping, is left to every man's discretion; yet they are not to abuse that interval

Zhou Zuoren,[5] a famous writer and thinker, was fascinated by the Japanese New Village movement. He visited *Atarashikimura* and published the article "*New Villages in Japan*" in 1919. In the same year, the Beijing student work-study mutual aid group was founded. Immediately after that, Mao published the article "The work of students", in which he mentioned similar movements in Russia, Japan, the United States, and the Philippines, and presented his own broader version of the New Village.

In Mao's vision, a student would work 4 h in agriculture, teach for 4 h r, study 4 h, and take leisure time (4 h). In this scheme, he not only put forward the "mind and manual education" idea similar to Kropotkin, but he also proposed a new lifestyle as "an independent and adult man", a "complete human being" as Kropotkin called it. Much more progressive than the New Village movement, Mao developed a completely new world from his educational village. The "man" and the "student" become the basic elements of a new family, a new school, and a new society—a society where everything is public: kindergartens, primary schools, other schools, libraries, banks, farms, factories, shops, theaters, hospitals, parks, and museums. During that time, Mao also hosted and participated in several "halftime studying and halftime working in the workshops" and "halftime studying and halftime working in the fields" groups.

He continued the New Village project while developing the socialist commune in 1958 and presented it in more detail in his important letter of instruction on July 5, 1966, to Lin Biao (五七指示). For Mao, the commune is the basic unit of society, a harmonious combination of industry, agriculture, commerce, education, and army. The commune members should be schooled as all-round laborers. In the letter, he explained that the army should be a "big school" where one should learn political affairs, and industry, and get involved in agricultural production. He also imagined applying this mix of activities to agriculture, industry, education, commerce, etc. Many units, especially the residential parts, were called "New Village" and are still called that in today's Chinese cities.

5.2.2 *Kropotkin, the Disurbanists, and Frank L. Wright*

Wright's combination of manual and intellectual work was seen as "borrowing an idea from the anarchist philosopher Kropotkin".[6] In 1899, Kropotkin moved to Chicago, where he lived in the Hull House commune where Wright often lectured.[7] Langmead and Johnson stated Wright's reading of Kropotkin's theories as a clear reference of

---

to luxury and idleness, but must employ it in some proper exercise, according to their various inclinations…".

[5] Zhou Zuoren was a Chinese writer, an essayist, and a translator, an important figure in the May Fourth Movement and the New Culture Movement in the beginning of the twentieth century. He is famous for translating many works of Japanese and ancient Greek literature into Chinese.

[6] See Wright (1935).

[7] See St. Clair, J. (2004).

his idea to link "between art and craft, design and execution, learning by doing, fair division of labor, the basis of admission, shared daily chores, repudiation of 'wage slavery', testimonial instead of diploma at completion, organization of vacations and financial implications of fees and products" and other issues.

The common foundation of Kropotkin's "Fields, Factories, and Workshops" and the disurbanist Okhitovich's[8] "Notes on Settlement Theories" is the dispersed production. In the article "Notes on Settlement Theories", one of the fundamental documents of the disurbanists, Okhitovich explicitly presents his awareness of Kropotkin's idea and agrees with it to a certain extent, but also offers some criticism through the comparison of the two works. For example, according to Kropotkin, the dispersion of energy (electricity) from the center to each part of the territory will encourage dispersal. This theory also inspired the utopia of Wright and many others. Okhitovich developed this principle even further into a centerless and ambient "common energy network" as the resources of the energy do not make a center. In this universal network which integrates even the smallest energy collectors "Each centre is the periphery, and the periphery is the centre of each item" (Okhitovich 1930: 14).

Of course, this concept of the energy network can be imagined easily as a universal principle of society concerning industrial production, schools, services, etc. For Kropotkin, the competition between large and small-scale production exists, and the production units must "mutually aid" each other to be competitive. In Okhitovich's world, this problem does not exist: the scale of the factory is territorial, and the network replaces the distribution centers—the production is manifold but is also one productive network. This concept is closer to the contemporary ideas of "Smart Grid, Dispersed Network Manufacturing", "Economy of Scope", and so on.

The influence of Wright on the Russian architects including the disurbanists can be traced back to the time before he was approached by Pravda in the early thirties before his dialogue with Arkin was published and eventually presented in 1937 at the All-Union Congress of Soviet Architects. While being a student at the Brera Academy in Milan, the disurbanist Ginzburg[9] expressed "an admiration for the work of Frank Lloyd Wright". (Ginzburg 1982: 12). He also drafted projects resembling Wright's houses. In 1927, he introduced Wright's Robie House in the magazine Sovremennaya Arkhitektura (Modern Architecture)—a magazine published by the

---

[8] Mikhail Okhitovich was a singular figure in Soviet architecture of the 20's and 30's, a Bolshevik sociologist, town planner, constructivist architecture theorist, and a member of the OSA Group (Organisation of Contemporary Architects) who constantly wrote in the OSA bulletin Sovremennaia Arkhitektura. He was the original developer of the theory of Disurbanism and was joined at a later stage by Moisei Ginzburg. As a supporter of the Left Opposition of Leon Trotsky, he was naturally at odds with the Communist Party during the Stalin era, was arrested, sent to the Gulag, and eventually executed in 1937.

[9] Moisei Ginzburg (1892–1946) was a famous Soviet constructivist architect, best known for his 1929 Narkomfin Building in Moscow. He is the founder of the OSA Group, and his book "Style and Epoch" (1924) was effectively the manifesto of constructivist architecture. He led the disurbanist school together with Okhitovich.

OSA group), which promoted disurbanism. Ginzburg continued to recommend and present Wright's work in the thirties.

5.2.3 A Common Agenda

One goal of the social agenda of the atlas is to eliminate two interconnected divisions—the disunion between the urban and rural and the division of labor[10]—through decentralization.

Kropotkin feels a society "must find the best means of combining agriculture with manufacture—the work in the field with a decentralized industry" (Kropotkin 1901: 6). His society is a place "where each individual is a producer of both manual and intellectual work; where each able-bodied human being is a worker, and where each labourer works both in the field and in the industrial workshop" (Kropotkin 1901: 5).

The disurbanists follow the same reasoning. Ginzburg in his letter responding to Le Corbusier states the impossibility to transfer the great number of peasants to the big cities without the destruction of the agriculture and the necessity of the elimination of "all the disparities between town and country". He recognizes very well the advantages of concentration in the development of culture, but also "the advantages of dispersal and decentralization for spreading culture as uniformly as possible over the population" (Ginzburg 1930).[11]

Miliutin is who believes that the Russian disurbanists and urbanists share the agenda of the elimination of the difference between the city and the country, but approach it in two different ways. However, in his idea, the "senseless" centralization of industrial production has to be eliminated to "settle the problem of the new redistribution of humanity" (Miliutin 1974: 60).

Mao, who admired Kropotkin's ideas, articulates a point of view very similar to that of the disurbanists, in which the "standard of living in the countryside must be the same as the one in the city, or better. This problem could be solved with the commune", and the peasants must be transformed into workers "in situ" (Mao 1961: 389–390).[12]

---

[10] Marx and Engels saw the separation of towns/villages (the urban and the rural) and the division of labor as two sides of the same coin: "Also characteristic of civilization is the establishment of a permanent opposition between town and country as the basis of the whole social division of labor" (Engels, 1942: 201).

[11] See Alexandre Tchayanov (1888–1937): *The Journey of my brother Alexei to the Land of the Peasant Utopia* (1920) about the difficulty to produce culture outside of the city.

[12] An almost identical agenda by Friedmann, who is very much influenced by Mao, was proposed in his "Agropolitan Development for Asia", "to transform the countryside by introducing and adapting elements of urbanism to specific rural settings. … transmute existing settlements into a hybrid form we call agropolis or city-in-the-field. In 'agropolitan development' the age-old conflict between town and countryside can be transcended" (Friedmann 1978: 183).

Wright has pushed this agenda to the extreme with his Broadacre city: "Elimination of cities and towns" (Wright 1935: 245). The urban and the rural are simply one: "the basis of the whole is general decentralization as an applied principle and architectural reintegration of all units into one fabric" (Wright 1935: 254). Every man will work in agriculture, industry, and the mental sectors, "to integrate general family small-garden and common little-farm production (to whatever extent this may be) and relate both to factory work and mental services a few hours each day" (Wright 1958: 169).

The ideas of the urbanists and disurbanists continue in the Program of the U.S.S.R Communist Party and are implemented in the Ideal Communist City. One can observe continuity in China's policy from the Maoist era, to the in-situ industrialization in the 80s, especially in Qian's idea of the sixth industrial revolution in 1984,[13] and today in Xi's ambition to abolish the urban–rural dualism and move toward integrated urban–rural growth.[14]

5.2.4 A Territory

The spatial agenda is embedded in the social agenda. As in the conclusion of "Fields, Factories, and Workshops", Kropotkin states the premise briefly and strongly: "to have the factory and workshops at the gates of your fields and gardens, and work in them" (Kropotkin 1901). If this figure is universally applied, this simple statement immediately and concretely generates an image of a territory.

This rapid overview can be considered as the starting point of an atlas of thoughts, which identifies and proposes a type of territory, which is necessary for the social agenda to function. Kropotkin described his territory by looking at Flanders, Belgium, the island of Jersey in England, and the Lyon region in France at the beginning of the twentieth century. This type is found also in the Maoist Commune construction

---

[13] Xuesen Qian (1911–2009) was a Chinese engineer who contributed to aerodynamics and rocket science and was one of the most famous scientists in China. As politician, he was also the vice president of the central committee of the Chinese People's Political Consultative Conference, and one of the leaders of the Party and the nation. His famous vision of the sixth industrial revolution refers to the knowledge-intensive agricultural industry, through which the difference between the urban and the rural, and between the peasants and the workers, is eliminated.

[14] This agenda of eliminating the division of urban and rural and the division of labor is shared by a much broader range of thinkers. It is reflected in More's Utopia, where each city is limited to 60,000 -96,000 inhabitants and the city itself produces agricultural produce. The inhabitants are sent in turn to the farms in the countryside, to work in the field. Everybody works in agriculture, and everybody also works as a craftsman or other professional. This vision is shared by the communists. In Marx's Manifesto of the Communist Party, the premise is "the combination of agriculture and manufacturing industries, the gradual abolition of the distinction between town and country, through a more equable distribution of the population over the countryside". Lenin, quoted a letter from the Russian disurbanist Ginsburg to Le Corbusier, stated clearly that "a resettlement of mankind is necessary, with the elimination of rural neglect and isolation and the unnatural crowding of huge masses into the big cities".

from the 50s to the 70s, the socialist city-territory proposed by the Department of the Socialist Settlement of the State Planning Committee (RSFSR), and the Broadacre City by Wright. It has a certain density of decentralized inhabitation, a decentralized mixture of agricultural and industrial production, and a universal decentralized infrastructure network. For Kropotkin, it was the universal supply of electrical power; for Okhitovich and Ginzburg, it was the common energy network and the railway and road network; for Wright, it was the motor car as the general mobilization of humankind, the intercommunication, and the standardized machine-shop production (Wright 1935: 345).[15,16] Many parts of China including the Yangtze River Delta belong to this category, with both continuities and discontinuities of the design themes and questions.

The territorial organization in the Atlas of Utopia

In the competition proposal of Okhitovich and other architects for the residential settlement in the industrial area of Magnitogorsk, eight residential belts are planned instead of one compact town. Each of those eight belts is about 25 km long, built along a motor road which has also public transport lanes that come together at the steel plant in the center. The residential belt meanders through the territory that has plenty of agricultural fields and natural features (e.g., forests, lakes, rivers). It uses those natural features to embed its cultural park. Along the belt, a variety of services and non-residential functions are planned. The city thus disappears together with the traditional villages, and both are replaced by a light but extremely extensive colonization of the territory. In the proposal for Moscow and The Green City by Okhitovich, Ginzburg, and other disurbanists, a similar linear organization of the settlement was presented to "break up" Moscow. The similarity is obvious in the territorial organization of the settlement and the detailed architectural design for each type of program in both competition entries.

After the influential article "Notes on settlements" (Okhitovich 1930) and the two competition projects, the RSFSR of which Okhitovich was a member started a more thorough elaboration of the territorial system implementation. A range of transitional schemes was set up according to the different contexts, such as the urban areas where the center of the district is the city, the industrial area where the largest processing company of raw materials, semi-finished production, and the agricultural or mining areas is in the center of the district (Fig. 5.2). The final scheme presents a universal and even dispersion of the settlements organized in a grid of triangles which represents the residential belts along motorways and public transport. Industries are located at each junction and have agricultural production as the background. The natural features, presented here as the river, cross the settlement system independently.

---

[15] The reflection on this type of territory continued in many parts of the world. To name just a few: the Zwischenstad concept by Sieverts in Germany, the *desakota* concept by McGee in Asia, and the Horizontal Metropolis concept by Bernardo Secchi and Paola Viganò.

[16] See Catherine Maumi on the relation of Broadacre with the economic crises (Maumi 2015).

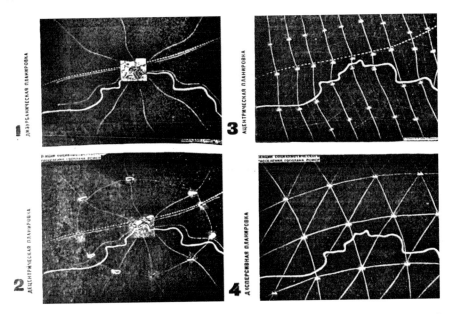

Fig. 5.2 Socialist City proposal by RSFSR, 1930. 1. Disurbanist planning of an urban area. 2. Decentralized planning of an industrial area. 3. Acentric planning, the area of agriculture or mining industry. 4. Dispersed planning, the consistent socialist city. *Source* Sovremennaya Arkhitektura, 1930 (6), page 2

From the reading of the layers of the Yangtze River Delta study area, a similar scheme can be designed. A mobility network grid, of motorways and canals, is evenly distributed over the territory. A relatively large industrial node is located at each junction. In the background is a complete agricultural layer. The large natural features scattered over the territory are fishponds, orchards, and tree nursery fields.

Density

In the words of Kropotkin, his utopian territory is with "a dense population, a large industrial area, and a sizable area of agriculture and horticulture". Under certain conditions (productivity, technology, etc.), a proper population density is necessary to provide a balance between production, consumption, and services. Around the beginning of the twentieth century, Kropotkin examined the density of territories such as Belgium in the 1900s (200 inhab./km^2), Saffelare in East Flanders (200 inhab./km^2), and the English island of Jersey as the densest (800 inhab./km^2). Broadacre City shows a borderless pattern with a density of 125–250 inhab./km^2, while the Maoist people's communes have an average of 10,000 to 20,000 inhabitants each. Several case studies of the Pearl River Delta suggest that each commune should be broken up into several settlements with a population of between 3000 and 5000 each. Friedmann's Agolopolitan District, inspired by the Maoist commune, has a density of at least 200 inhab. /km^2.

Today's Yangtze River Delta has a dispersed population density on another scale. The Tai Lake watershed, as the main part of the Yangtze River Delta, has a population density of more than 1600 inhab./km^2 on a territory of 36,900 km^2, which is about 3 times the size of Flanders (13,522 km^2) and more than 3 times its density (477 inhab./km^2), and close to the density of Leuven (1800 inhab./km^2), a medium size city in Europe. Even in its "third space", Tai Lake's hinterland without any large urban cores, the density is more than 800 inhab./km^2 (in the 50 × 50 km study frame), which is comparable to medium Belgian cities such as Kortrijk (940 inhab./km2) and Hasselt (750 inhab./km^2). Besides the large water features, hills, agricultural land, and other uninhabited spaces, the population is spread over many areas with a higher density. This does not include the possible future immigrants.

This density is critical: it is an urban density. One of the most criticized aspects of decentralized and diffuse urbanization, or one of the most convenient arguments for compact urbanization, is the efficiency of facilities and infrastructure. The low density presents a challenge for the economic organization of kindergartens, schools, care centers, hospitals, etc., as seen in many parts of Europe today. In the Yangtze River Delta, the high density provides a unique opportunity. For example, in Tangqi, the population density is higher than 1200/km^2 in many villages. It means roughly 120 students for an elementary school (a class of 20 for each grade, 6 grades in total) and 120 students for a secondary school within a radius of 1.5-2 km distance of 20 min on foot or 10 min by bike. The density offers the possibility to organize small schools, whose students can commute on foot—an incredible advantage that is hard to imagine in other dispersed urbanities. A similar conclusion can be drawn for public transport, medical services, etc.

On such a territory, every aspect of life is imagined in the atlas: production, education, living, leisure, energy production, etc., and the space for each of them is illustrated. The following paragraph takes the school system as an example of imagining a new territorial system from and for this specific type of territory.

5.2.5   *The Schools: an Element of the Common Agenda*

A vision of education is a necessary component of the objective of eliminating the division of labor. Kropotkin, Wright, the disurbanists, and Mao share the same criticism on the specialization of knowledge. They promote the idea of a complete education of intellectual and manual knowledge, in other words, the combination of education and production. According to Kropotkin, we should "maintain that in the interests of both science and industry, as well as of society as a whole, every human being, regardless of their background, ought to receive an education that would enable him, or her, to combine a thorough knowledge of science with a thorough knowledge of handicraft. We fully recognize the necessity of specialization of knowledge, but we maintain that specialization must follow general education, and that general education must be given in science and handicraft alike. To the division of society into brain workers and manual workers, we oppose the combination of both kinds of activities;

and instead of 'technical education', which means the maintenance of the present division between brain work and manual work, we advocate integrated education, or complete education, which means the disappearance of that pernicious distinction" (Kropotkin 1901, p. 188). This "complete education" of Kropotkin is provided until the age of eighteen or twenty, like the Ginzburg and Okhitovich idea that "specialization in secondary school is dangerous. Polytechnic secondary education is needed", and education has to be done through production even from childhood: "they (the children) are accustomed from an early age to the three most important things: independence, sport, and production processes" (Ginzburg and Bard 1930).[17] Mao, who had reflected on this type of education since his youth and was influenced by the Soviet Union, formulated it as the guideline for education in a speech in 1958: "Education must serve proletarian politics and must be combined with production; the laborers should be intellectualized and the intellectuals should be laborized" (Mao 2008: 291). Similarly, Wright proclaimed that "specializations should no longer be so much encouraged…Then, general education by and for life in the free city should be had by *doing*…" (Wight 1958: 205).[18]

This social agenda promptly dictates a functional agenda: the combination of the school and the production space. In 1958, Mao specified this vision as the main model of education:

> If possible, every middle technological school should try to run a factory or cultivate the land for farming, to be self-sufficient or semi-self-sufficient. The student should work while studying. …All the middle and primary schools should sign a contract with local agricultural cooperatives, to participate in agriculture and agro-industries. (Mao 1958: 64).

The Maoist government intended to implement this agenda on a large scale. As Liu proposed, the half-work, the half-study school should be the main form of education in China.[19] According to the 1961 Regulation Of The Work In Rural Communes (Draft), each secondary school had to be transformed into an after-work-hours school, and the students including the children were officially registered as laborers. If one-day work of a male laborer is counted as 12 *gongfen* (a unit according to which the

---

[17] After the October Revolution, the combination of education and production, even production as the base of education, has always been the guideline of education in the Soviet Union.

[18] This type of education, as a necessity for the elimination of the division of labor, is also shared by many other thinkers. In More's Utopia, each man or woman must learn from school and from practice, to learn agriculture and some peculiar trade, without excluding the higher study and research.. Marx also notes the necessity of the combination of education and production: "…an education that will, in the case of every child over a given age, combine productive labor with instruction and gymnastics, not only as one of the methods of adding to the efficiency of production, but as the only method of producing fully developed human beings" (Marx, 1867: Volume 1, 529–530). However, his principle was not shared by all the communist practice. See the chapter "The Basic Aims of Soviet Educational Policy" (De Witt 1961: 5), which shows the conflict between Marx's idea of education that "productive physical labor should be an integral part of the educational process with book learning" (De Witt 1961: 6) and the *functional education* in the U.S.S.R whereby each individual is to be equipped with specific knowledge and skills subordinated to the need of the state. This idea of functional education was also continued in *The Ideal Communist City*.

[19] See "教育革命"的历史考察: 1966–1976, by 陈晋宽page 140 二, "半工 (农) 半读"与"两种教育制度".

salary is distributed), the one of a strong male student is counted as 9–11 *gongfen* and the one of a child as 3 *gongfen*. This practice was not a pure utopian dogmatic communist idea, but more, besides the economic reasons, a reflection on the transformation in the countryside where the rural industry was developing, and the rural individual often became a farmer and a worker.[20] The new education system was encouraged to deliver a new type of laborer who could carry out different types of work according to the needs of the society and of his/her interest, just as the decentralized European industry in the early twentieth century also contextualized the idea of Kropotkin of combining the intellectual and mutual education.

In the idea of the disurbanists, the space of combination of education and production merges into laboratories where the high-tech equipment is: "…laboratories and equipment are not only the school but also the purely productive. Students not only learn, but also carry out socially useful productive work" (Ginzburg and Okhitovich 1930).[21]

The combination of education and production urges education to be organized according to the decentralized industrial system in the territory, which means a new system of schools from its territorial organization to its architectural design. Instead of being passively subordinated by the organization of industry, this new school system was imagined to radically and extensively utilize this special condition of the territory.

In the "Green City" competition entry for Moscow in 1930, Ginzburg and Okhitovich proposed such a school system. The first level is a decentralized nursery and children's care, a group of buildings with 15–20 kids in each building, which are located along and connected to the residential belt—an endless and continuous row houses on the territory. The decentralization could prevent infection, but also provide space for children of four to five years old to start cultivating small plots, take care of the birds, get acquainted with the animals in the zoo, or identify vegetation in the botanical garden. The second level is the decentralized schools, each designed for only 150 students, and located close to the collective hostels in forests or glades where the students sleep in common dormitories and spend all their time in a group of peers. Each school can teach many general subjects (reading and writing, mathematics, social studies) without needing special equipment. In the third level, the element of the production is added to each school. First, the school specializes in a certain category of knowledge: natural science, agriculture, forestry, social sciences, industry, housing, construction, etc. The specialty of each school is designated by

---

[20] In the Party document "Education Must Be Combined with Productive Work" (1958), this situation was described as evidence of the beginning of a communist society.

[21] Miliutin even pushes this agenda further: the combination of education and production is not just for the children and adolescents, but also for the workers: "By the unification of educational institutions with production laboratories, workshops, fields, libraries, archives, etc., we will not only achieve significant economies, but will also make possible the great idea of turning an industry into a school. Every male and female worker (including the cook in factory kitchens, the hospital attendant, the courier in Soviet institutions, the shepherd on the *sovkhoz*) will have the opportunity to become an engineer, surgeon, economist, agronomist, etc., during his or her usual work" (Miliutin, 1930).

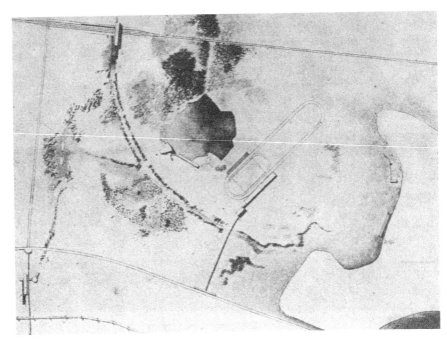

Fig. 5.3 Auditorium as part of the school system spread in nature. Sotsgorod: The Problem of Building Socialist Cities, By N. A. Miliutin Edited by George R. Collins, William Alex, 1975. The MIT Press. https://doi.org/10.7551/mitpress/6353.001.0001

the specificity of its territorial location; for instance, a school located in an agrarian area will be specialized in agriculture. Second, each school receives laboratories and equipment for its specialty, which are not only educational but also purely productive. Students not only learn but also carry out socially useful productive work. The polytechnic learning circle, the idea of a complete education, is carried out by all decentralized schools on the territory. The diffuse mobility network, which connects all the schools, enables the students to periodically receive training in other schools. This movement of the students on a vast productive territory provides a tremendous pedagogical value in terms of extending their vision and understanding of geography. Each school has an auditorium that functions as a civic center for the entire community, where not only lectures but also films, shows, and concerts could take place—a connection between decentralized territorial education and decentralized territorial living (Fig. 5.3).

The schools in Broadacre City are designed the same way: they are 1-floor buildings for about twenty-five students, decentralized in parks in the countryside, providing large playgrounds and flower and vegetable beds for each pupil to work on. The higher form of education is the design center, a productive, and educational apparatus at the same time—a twin of the specialized schools in the "Green City". This type of school is planned also in the countryside, so the students could work 3 h a day in the field to earn their living, like the students in Mao's New Village imagination.

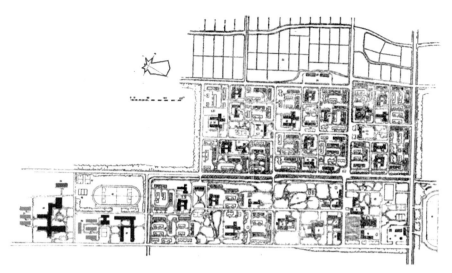

Fig. 5.4 Masterplan of a residential settlement of Shantou Waisha people's commune. (Architecture Department, South China University of Technology 1959). Dark red: the elementary school and the secondary school. Light red: the collective dormitory of the students. Bright red: the factory of the elementary and secondary school. Elaborated by author

Similarities can be found in the Maoist program. According to the Architecture Planning and Design of People's Commune (Architecture Department, South China University of Technology 1959), the scale of kindergarten should be small (75–125 children), and each group should be composed of about 20–25 children. The plan of a kindergarten contains a group of several small buildings, with plenty of open space around. There should be space for the tree and/or flower nurseries and for keeping animals in kindergarten within the kindergarten. In the masterplan of a residential settlement—one of the many spreads throughout the territory of Shantou Waisha people's commune in 1959, a mix of residential, public buildings, services, production, commercial space, and schools can be found in the lower part of the plan, which is reminiscent of the proposal for Magnitogorsk by Leonidov. Nature and water features surround each building. On the western end of the strip, the elementary and secondary schools are located, which have an agricultural field on both sides. A factory is dedicated to and shared by the two schools. Within the strip, sports fields and swimming pools are serving both the inhabitants and the students (Fig. 5.4).

The theme of the combination of agricultural production and education continues in today's practices. For example, Secchi and Viganò propose "Garden in The West" as one of the strategic spaces for Brussels 2040, where the clusters of schools are combined with the agricultural land in the western part of Brussels (Fig. 5.5).

The system of the school, extracted from the work of different architects and thinkers, is an example of a complete spatial agenda that functions in its social agenda and its territory. More than just a way of thinking or a methodology to imagine a system, constrained by its historical and territorial limits, this system is not fully

Fig. 5.5 Model photograph: Strategic Space: Garden in The West. The school is strategically located as the entrance to fields as agricultural productive space and public space. *Source* Vision for Brussels 2040, Studio Bernardo Secchi Paola Viganò with Creat, Egis Mobilité, TU München and Ingenieurbüro Hausladen GMBH, Karbon', IDEA Consult

implemented. It is presented here as a legitimate option for the future for a similar setup in the Yangtze River Delta for instance. The Delta does not only share the similarities in the physicality of the territory that the atlas is dealing with, nor the similarities in the challenges of the urbanization process, or the similarities in the social agenda—especially the communist agenda still present in China today, but it also provides conditions—such as the higher density, more advanced technology and industrial system, etc.—that are even more suitable for the system to be implemented.

5.3 Imagining the New Systems

Imagining the new systems is an attempt to reflect not on the main issues in today's urbanization but on a specific territory (Fig. 5.6): the third space of the Yangtze River Delta. Many of the issues are common to what the utopians confronted: the mix of agricultural and industrial production, the size and distribution of facilities and activities, the ways to organize mobility and infrastructure, education, etc. Others are only increasingly being examined today, such as ecosystems, migration, and pollution. Although the context was completely different in the era of the utopians, the main issues remain: how to put an end to the urban–rural divide.

There is no doubt that some of the visions of the utopians are obsolete, and the purpose of this chapter is not to praise the wisdom embodied in their work and implement it in a territory that is completely foreign to them. Nevertheless, there are two aspects of their work that draw attention. The first is to eliminate the urban–rural divide and set aside the differences and even contradictions between the utopians, the territory needs to be organized into decentralized and horizontal forms. Those forms are exemplified by the vegetable gardens of Kropotkin, the communes of Mao,

Fig. 5.6 Detailed model, about 4 m by 4 m, is designed by the author and constructed by the author with the help of students from EPFL, Lausanne. The model is a presentation of the imagined "utopia" on the site of Tangqi, China. The model takes the Broadacre City model as a reference, in terms of size, scale, and level of detail. The model and a part of the work of this thesis were presented in the exhibition of "Horizontal Metropolis" at Archizoom-EPFL Lausanne in 2015 and "Horizontal Metropolis—a radical project" for the 15th International Architecture Exhibition—La Biennale di Venezia in 2016. Photograph by author

the energy network of Okhitovich, the linear housing of Ginzburg, the "Broadacres" of Wright, and many others. The third space of the Yangtze River Delta is to a large extent such a territory. Decentralization and horizontality, as described in the previous chapters, are deeply rooted in the territory's historical construction and reinforced in many recent developments.

The second, an esthetical exercise by the utopians, both the architects and non-architects, is a clarification of certain common and specific spatial premises as an intermediary step, separate from the grand ideology behind and the concrete solutions in the forefront. Those aspects illustrate the purpose of eliminating the urban–rural divide, not through abstract numbers but perceivable life qualities, and can be extremely simple but powerful at the same time. For example, "every room should have cross ventilation, open views (on both sides), sufficient volume, and ideal (natural) lighting", a simple and sensible premise put forward by the Russian disurbanists,[22] but at the same time almost unimaginable both in today's over-congested urbanization and declining rural areas. Behind this simple issue is radical planning of the space, which has to accommodate views, wind, and light. More than that, it even suggests looking for certain territories suitable for this type of planning. In today's China, the construction of such a program can be an effective way start to a discussion that goes beyond the overly abstract and ideological ambitions and even beyond the overly detailed and individualized projects. It can be understood easily by and involve officials, designers, and inhabitants. Furthermore, this discussion can establish quality details without pointing to a specific solution.

The quality outlined by the programs is imagined not through individual elements but through systems, because the disurbanists' linear settlement is a system in which the row houses, the mobility lines, the energy supply, and the landscape are all organically organized. Through such an organization, the specificity of the territory is shaped. The utopian projects, such as the Broadacre City in Boston and the Green City in Moscow, are seen here as a clarification of those programs rather than the only design solutions. Moreover, the evident differences in the utopians' design projects prove that the common agenda can produce great variety in design depending on the context.

Imagining new systems is an exercise that attempts to connect three elements: the territory and its society, the proposed designs, and the unique opportunities provided by the conditions and the physical environment of the territory itself.

---

[22] In the competition entry of the Green City Moscow in 1930, the disurbanists claimed "it is extremely important for a tenant not to have any structures in front of his eyes on both sides along the lines of settlement", and one of the main advantages of the linear housing proposal is hygienic.

With the usual urban resettlement, urban blocks are built; that is, they are nothing more than closed or semi-closed air bags of greater or lesser magnitude, depriving residents of these quarters of ventilation, open horizons, spaciousness, and ideal illumination. We are striving for such a settlement system, in which all these shortcomings would be eliminated. This organization generates tremendous views of forest and agriculture on both sides of each room:

From the line of the road, retreating a minimum of 200–250 m along a continuous park strip, there are ribbons of residential buildings. Thus, each link in the housing has one side of the park strip at 400–500 m width in front, and on the other the vast expanses of green massifs of forests and fields.

The territory and its society, as discussed in the previous chapters, include the problems of demographic growth and the increasing number of migrants, the decline of agricultural production and the ecosystem, the demand for public spaces and facilities, the lack of sustainable mobility and energy supply, the transformation of industrial production, and the superfluous housing construction. Those issues are affected in a cross-cutting way by the five systems presented in the following paragraphs. For example, the issue of demographic growth is reflected in the system of schools, energy, and housing, and the public space is reflected in the road and waterfront systems.

Taking the work of the utopians as a reference, my intentions for this territory are proposed. Some of these are taken over from the utopians because they are still valid to resolve some of the current issues. The others are invented, inspired by the specificity of the territory. For example, the territory with dispersed urbanization is often accused of depending heavily on private vehicles. In my proposal, "every elementary school student should be able to go to school on foot safely though a green space" is a simple premise that refers to the safety of the students, which reveals the potential of this type of territory that is often neglected.

These design issues are presented by the five systems: fishponds, orchards, and schools; the asphalt space; the concrete space and the waterfront; industry and the energy network; and housing. Imagining new systems for this territory is not an attempt to come up with concrete solutions for specific issues in the future, for example, the future organization of mobility, industry, or housing, especially at a time when the future of those issues is extremely open. Rather, imagining these systems is an attempt to show how certain premises are realized by the specific conditions and space of the territory. Compared to other territories, some of the conditions that are critical to certain design issues are unique in the third space of the Yangtze River Delta, such as the high population density and the existence of a dispersed forest. Imagining these systems, combined with the space within which they are to be realized, creates certain forms or intangible figures. In between the extreme and almost shapeless dispersion and high concentration, which can both be found in the territory, the figures provide an intermediary scale or even an alternative way of tackling the question of the ideal degree of concentration/decentralization, the ideal degree of hierarchy/horizontality, and the ideal degree of centeredness/isotropy.

5.3.1 Fishponds, Orchards, and Schools

The fishponds and orchards form a continuous public space—a large-scale figure whose dimension has thus far not been sufficiently recognized. This public space will not function as a space for production, recreation, and nature only, but also as a space for public functions and movements (Fig. 5.7). It will be equipped with a system of paths that serve as a slow mobility system connecting all metropolitan

|0 | | | 3.2km/2miles|

Fig. 5.7 Imagination: schools and other public buildings spread in the fields of fishponds and orchards. Illustrated by the author. Red: Schools and other public buildings; black: public space; pink: orchards; dark gray: fishponds; and light gray: water

programs, civic buildings, schools, sports facilities, public transport nodes, industrial estates, and eventually the backyard of each house. Pedestrians and cyclists will be able to move though this intense path system in the park without being hindered by cars.

Fishpond

The use of the fishponds will be diverse. Several abandoned fishponds, which are located too far from housing, will be transformed into wetlands to purify water.[23]

---

[23] The fishpond to wetland operation is already being practiced in the Delta, including the largest urban wetland park in Suzhou—the Huqiu Wetland Park—which was transformed from fishponds.

The wetland with its vegetation would bring diversity to the landscape. The banks of the fishponds will be soft whenever possible, to improve biodiversity. Some of the fishponds will be transformed into open water ecological swimming pools, with naturally purified water. The other fishponds will be maintained productively, as part of the traditional orchards-dike-fishpond model.

Orchards

Because of the increasing population, more orchards are to be created, to connect the currently fragmented plots into a continuous space with trees. The orchards could be extended into other spaces, for example, the courtyards of the schools and kindergartens, between the sports fields, and into the industrial estates. A diversity of orchard types such as grapes, peaches, and oranges could be introduced alongside loquats.

School

The school system is similar to the one presented in the *atlas of utopia*. The population density provides an even more suitable condition for this system to be implemented. Besides combining intellectual and physical knowledge, education, and production, the social agenda can be extended: could we imagine a system of schools in which the students could safely walk to school, to the auditoriums, to the production sites, without having to interact with cars? Could we imagine that the students play in the sports fields and playgrounds with nature within walking distance? Could we imagine that every classroom has quiet surroundings and a view (Fig. 5.8)?

In our study area of ±10 km², the population is expected to grow to 20,000. It means 2000 students for elementary school and 2000 students for high school. Taking the elementary school as an example, the 2000 students are divided into 80 classes, and all 6–8 classes form a unit with students from 3–4 different grades. Each unit is within 300 m of a public transport node and connected by the path system in the orchards, which allows the students to walk from their homes to schools in less than 20 min. Therefore, the schools also function as civic spaces and as entrances to the orchards. Each unit is surrounded by agricultural fields, orchards, fishponds, and vegetable gardens. Fully equipped light-building structures are set up in the orchards, next to the fishponds, and in the middle of the vegetable gardens. These function as classrooms for lessons, discussions, or self-study, but also as tea rooms, bars, and coworking spaces—spaces to be shared by the children and the adults. Terraces are developed similarly, to open-air classrooms. The sports fields and playgrounds can be found throughout the fields and orchards.

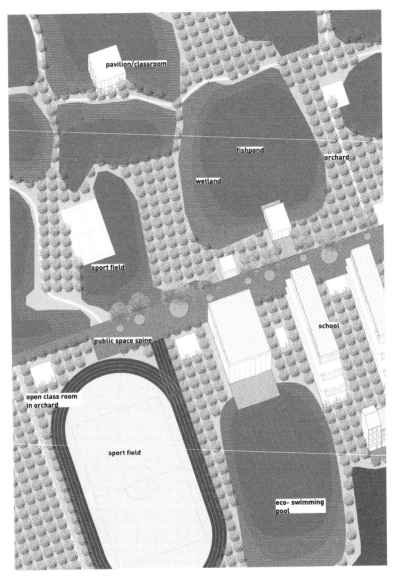

Fig. 5.8 Imagination: a school system spread in and connected by the space of orchards and fishponds. Illustrated by the author

The students work in agricultural production as part of their education. The study, design, and maintenance of the orchards, fishponds, dikes, water management, and other objects needed to operate this productive landscape are partly their responsibility. With modern communication technology, the exercises of the students could be connected to those of other local, regional, and even global institutions/agents: kindergartens, schools, research stations, peasants, etc., to tackle shared environmental and ecological problems today. The students also work in industrial production as part of their education. The decentralized industrial production and networked manufacturing enable the students to be involved not only as operators of the instruments, but also as participants in the research, design, manufacture, and other parts of the chain. They could also utilize the industrial platform to develop their own products. The workers and the students share the laboratories. The auditorium, just as the disurbanists imagined, is part of the school system but also used for lectures, cinema, and other public programs, a place shared by the children and the adults.

Within the territory, the students can use public transport, but they can also walk or cycle via the orchard system. The movement itself, passing every element of the territory, becomes a part of their education. The entirety of the diverse programs, connected by the environment of orchards and the territory itself and the movement through it, is the complete school.

Public buildings

Metropolitan features, such as the botanical garden, the auditorium, and the stadium, are located next to the largest natural figure—the lake. The open water is visible and accessible via water mobility and other public transport. More isolated places, for example, the islands or small peninsulas, are suitable for research centers, care homes, hostels, and temples.

A Park

The orchards, the different types of water features, the paths, the pavilions, the playgrounds, and the sports fields create a park for a myriad of activities: production, schools, a library, and a club. All are constantly using the park. It provides a unique space that is crucial for these programs to function as part of the social scheme that is imagined (Figs. 5.9 and 5.10).

Wright's Broadacre City proposes such a "park" space, a green natural area that is not specifically described in his text, a figure occupying more than 1/3 of the drawing of the Broadacre City that organizes a special group of elements. This figure is presented even more radically in an earlier sketch by Wright, in which there are natural movements that strongly contrast with the grid of the Broadacres. Many elements are located inside this figure and naturally follow the movement of the figure, instead of the orthogonal configuration in the later versions of the sketch. There are 51 elements listed in the Broadacre City drawing, and 24 are connected to or located within the natural figure. Most are elements related to culture, education, and sports, whose functionality depends on this natural environment instead of on

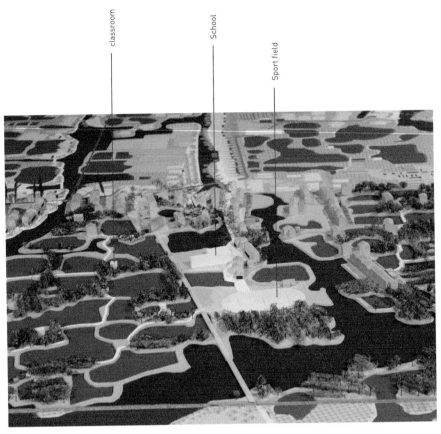

classroom

School

Sport field

Fig. 5.9 Model view of the park. Photograph by author, model by author in collaboration with Lab-U, EPFL

the Broadacres. One could also find the similarity between this park and Tschumi's Parc La Villette; in both, the program (the school in the case of Tangqi) is split into small units colonizing the entire park and being adapted according to the context of each part of the park.

The park does not attempt to bring back any nostalgic image of the public space in terms of architecture, material, and use. It tries to introduce a layer of contemporary activities onto the continuity of the form of the territory, to discover qualities from the existing model for the new setup.

Hostel

Fig. 5.10 Model view of the park. Photograph by author, model by author in collaboration with Lab-U, EPFL

5.3.2 The Asphalt Space

The space of asphalt, or the space along and related to the county roads, is the basis of a territorial mobility network in the third space and provides the highest connectivity in this space, as well as being the space for public transport (Fig. 5.11). The asphalt space runs through both the agricultural fields and the built areas. It includes the fir trees, the forecourts of the small industries, other activities along the roads, and the entrances to the industrial estates and the small residential roads—the concrete surface. Those elements enable the asphalt space to be transformed into an "urban" boulevard in this low density but universally inhabited urbanity.

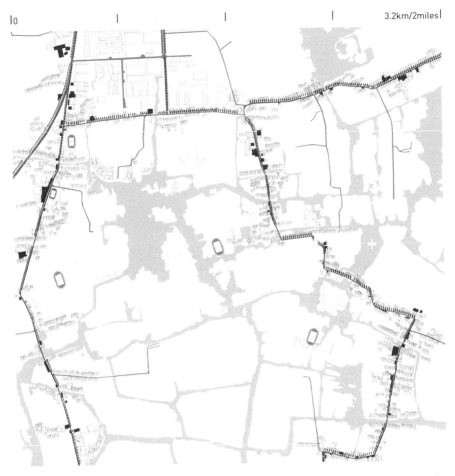

Fig. 5.11 Imagination: the space of asphalt as a space for public mobility, iconic trees on two sides, and transformation of industrial production. Illustrated by the author. Red buildings: industry + public facilities + activities along the asphalt space. Red line: public transport. Dark gray: space of asphalt. Light gray: water surface

The continuous component

Currently, the road is ±6.5–8 m wide, excluding the fir trees on both sides. The space for trees is minimal. A continuous space for mobility (a road) of 6.5 m wide will be built in the middle, and the rest of the surface serves as the enlargement of the space for the trees. A bike path of 2.5 m wide is added to the existing section of the road, outside of the tree spaces. The bike path, the fir trees, their space, and the 6.5 m surface for mobility (the road) are the continuous component of the road, which provide strong legitimacy and identity.

Public transport and private vehicles will share the continuous 6.5 m space. The first step will be a universal implementation of bus lines on the 6.5 m linear surfaces,

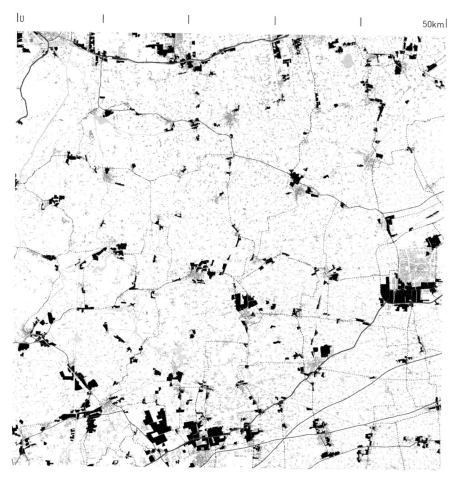

Fig. 5.12 Territorial system of public transport and industrial nodes/urban centers. Illustrated by the author. Red line: public transport. Black: industrial nodes / urban centers. Light gray: buildings (source: Tianditu, National Catalogue Service for Geographic Information. Elaborated by Author.)

which is currently the case in some parts of the territory, to provide the basic layer of public transport (Fig. 5.12). The bus is still the most suitable public transport model for this population density (± 1250/km^2). When this density is around 2000/ km^2, a tram can be considered, which together with a cargo tram will significantly reduce the use of private vehicles. In the future, a more radical no-car scenario can be envisaged: private car ownership will almost disappear. The different ways of car-sharing and the use of public transport will be organized much more efficiently through artificial intelligence, which may reduce the conflict between public transport and cars. Some of the industrial platforms, which are connected to higher mobility arteries and canals, could function as logistic nodes from where smaller trucks or cargo trams could distribute the goods via the county roads. The reduction in the number and size of the trucks, and the optimization of their performance and schedule, could significantly mitigate their impact on the country roads.

High-intensity area

In some parts of the territory, the asphalt space runs through the villages and the towns, with buildings on both sides. At least 2 m of sidewalk will be provided on both sides of the road. The sidewalk is made of permeable bricks that are recycled from local construction waste (broken-up concrete roads, bricks, and concrete slabs from demolished old buildings, etc.). The sidewalks will be at the same level as the rest of the asphalt space. Gutters will be provided on both sides.

Because of the expected increase in population in the future, civic buildings, schools, and commercial space will be needed. Many industrial production spaces along the road could be transformed to accommodate these functions. Transparent materials will be used to take advantage of the urban and rural views on both sides of the buildings. Small plazas will be built in front, to create a series of urban nodes. Their light and vast roofs could be ideal places for solar panels. The section is interrupted at some nodes for a public transport stop, the forecourt of important civic buildings, a bridge, etc. At the nodes, exceptional materials, such as bricks or stones, could be used. Orchards, along with perpendicular roads, will be extended to the asphalt space as an entrance to the villages.

Gradually, the road could be altered. Especially in the densely populated parts, a common material, such as bricks, could be used for the entire surface, like the pedestrian walkways in today's cities. The urban typology of space could be introduced in the third space.

Low-intensity area

In other parts of the territory, the asphalt space runs on one or both sides of the agricultural fields. The tall fir trees make the asphalt space visible from the open field. When the asphalt space runs through the orchards or the woods, no fir trees are planted on the sides.

When there are no buildings on the side of the asphalt space, a *wadi* (natural ditch) will be constructed—a lower space wide enough (minimum 3 m) to provide considerable water storage capacity and accommodate different types of vegetation. The *wadi* could be integrated also into the field drainage system next to the asphalt space. The rainwater in the high-intensity areas will be led directly to the gutters, which are connected to the *wadi*.

A linear public space: the boulevard

The linear space—the roads—is a crucial layer in Broadacre City. Besides its role in guaranteeing the functionality of a group of other elements, most of which are related to production, the fluidity of the movement itself is a fundamental quality of Broadacre City. Special elements are designed not only for this fluidity, such as the special highway intersection but also for the experience of such movement on this open territory, such as the trees that are planted in rows perpendicular to the roads (Fig. 5.13).

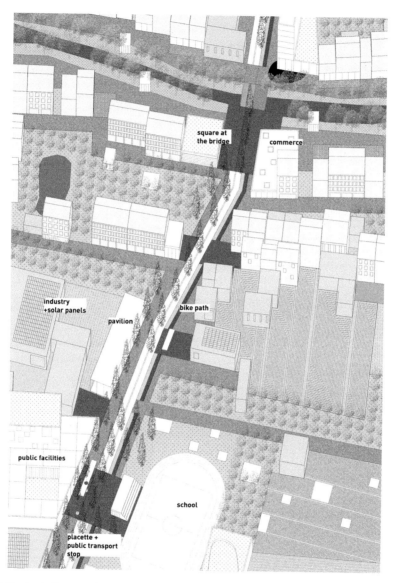

Fig. 5.13 Asphalt space of a county road is transformed into an urban boulevard, creating a series of public spaces. Illustrated by the author

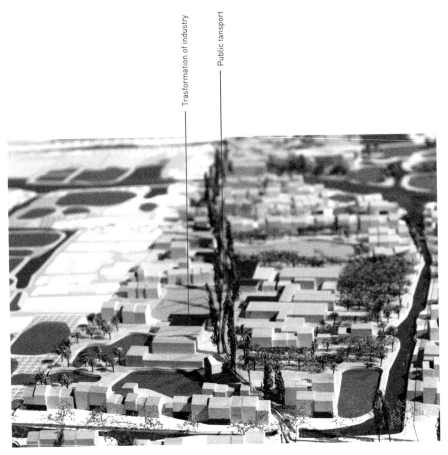

Trasformation of industry

Public tansport

Fig. 5.14 View of the asphalt road as an urban boulevard. Model by author in collaboration with Lab-U, EPFL. Photograph by author

In the third space, the road has a different movement. It runs into both the urbanized area and the open fields. Some parts of it demand/reject fluidity. Meanwhile, its functional consistency as a county road contrasts with its fragmented spatial appearance, which was historically far more evident.

The third space lacks articulation of space, in terms of public and private, the material, and the typologies. While squares, boulevards, and bike lanes are consistently built in urban environments, those elements are rarely seen in the third space. At the same time, the road in the third space, especially in relatively densely populated areas, is publicly and informally occupied by markets, terraces for restaurants, meeting points, etc. The imagined concept attempts to articulate the asphalt space, from the homogeneity of function (mobility) and material (asphalt) into a sequence of diverse public spaces: an asphalt road in between the fir trees in the open fields, a road running through the orchards, an asphalt road with large brick sidewalks in

Abies trees

Fig. 5.15 View of the asphalt road as an urban boulevard. Model by author in collaboration with Lab-U, EPFL. Photograph by author

the villages, and a brick-paved square covering the entire school forecourt which has camphor and peach trees and can double as a market from time to time. A wooden terrace is set up when there is a restaurant, a concrete platform located under the fir trees serves as a bus stop, a wide bridge as a viewing point over the water...It is a space, in its scale and function, equivalent to the boulevard in the urban context. When this space runs through a city, it is an urban boulevard (Figs. 5.14 and 5.15).

Instead of continuing the current popular construction of squares, temples, and cultural centers, which intends to reintroduce the ancient program and the architectural language, the transformation of the asphalt space as the starting point to construct public space attempts to capture the actual dynamics of the third space with the road as the central civic space. It not only answers the need for a locally articulated public space but could also potentially be a space shared by the urban and the rural, a space repeated as a continuous component throughout the territory and diverse in other parts according to the local contexts.

5.3.3 The Concrete Space and the Waterfront

Today, the concrete surface is mostly at the perimeter of each *yu* and has a wider Sect. (6–8 m) in the east–west direction also including the forecourt of the peasant houses. These houses, historically and today, are located along these concrete surfaces. Thus, the linear concrete surface functions as the "residential street" which is always a waterfront—a unique quality in the territory. These wide linear concrete surfaces (6–8 m) along the different *yu* compose a continuous concrete belt interrupted by water between the *yu* (Fig. 5.16).

Today, this space is used for travel, parking, washing, and drying agricultural produce and clothes. In more densely populated areas, this space is already the main

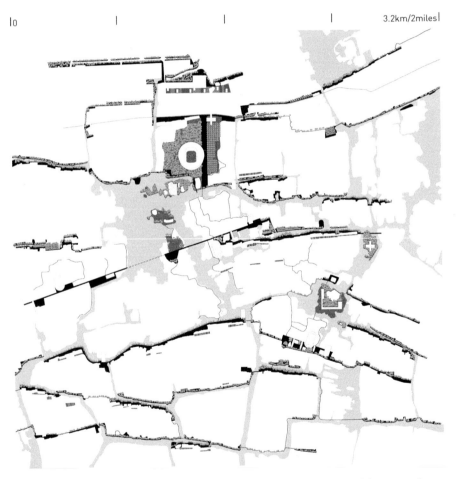

|0 | | 3.2km/2miles|

Fig. 5.16 Imagination: the space of concrete along the water is transformed into a continuous public waterfront. Illustrated by the author. Gold: vegetations along the waterfront. Black: public space along the waterfront. Orange: collective gardens. Light gray: water

public space in the neighborhood, despite its basic spatial performance. However, there still is a lack of public space, especially sports facilities and playgrounds.

In this scheme, this concrete surface is connected by bridges and requalified as a continuous and facilitated public waterfront. The continuity can connect a critical number of houses and inhabitants, thus generating enough flow for facilities and other activities (Fig. 5.17).

The common components

Informally constructed volumes (toilets, warehouses, kitchens, etc.) will be removed and reorganized, and the concrete surfaces will be gradually replaced with permeable bricks recycled from local construction waste, the same as the sidewalk of the asphalt road. The change of material will create a pedestrian atmosphere where the use of the car is discouraged. A mixture of camphor, peach, and willow trees, which are indigenous and widely planted in the villages, will be used for this linear space. The flowers and diverse trees create an intimate atmosphere for the neighborhood in contrast with the tall fir trees along the asphalt road. On some wider plots, small sports fields, playgrounds, and squares could be set up, with access points to the water, stairs, terraces, and balconies. The enlarged brick space, the types of vegetation, and the spread of facilities are the common components of all sections of the waterfront.

Bridges

East–west bridges will connect the different *yu* and the different patches of concrete surfaces to create a continuous belt. North–south bridges will be built between the different belts. The bridges are for pedestrians and cyclists only and function as a public space, as was the case throughout history. They become small public nodes along the concrete belt, where exceptional features could be developed such as a small plaza, high-rise residential buildings, neighborhood centers, hostels, and community centers. The water buses have their stop by the bridge (Fig. 5.18).

Ground floor

The houses along the waterfront should have open forecourts toward the water, as in former times. Each house can have a forecourt/garden of a maximum of 3 m with minimal/low fences only. The ground floor will be used for miscellaneous activities such as workshops, commerce, restaurants/cafes, housing, and public facilities, needed by the future inhabitants. The waterfront surface could be used temporarily for tables and chairs, and as an informal playground for kids, and vendors.

Kindergartens

Kindergartens are one of the public programs to be implanted on the ground floor of the houses, to make it easy for the parents to reach their children. The kindergarten is a transparent volume connecting the waterfront and the agricultural lands/orchards/

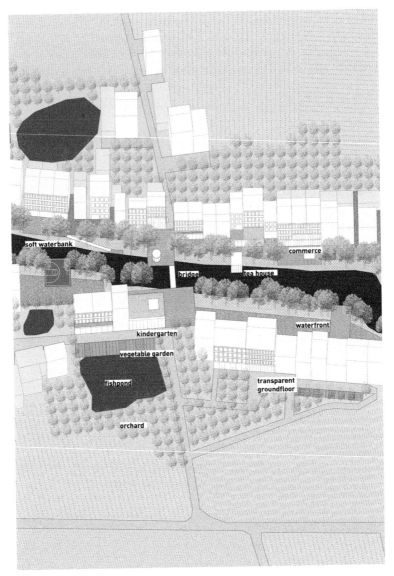

Fig. 5.17 Imagination: the space of concrete along the water is transformed into a continuous public waterfront. Illustrated by the author

fishponds/gardens at the back of the building. The kids could learn from the fields and get the basic knowledge about soil, water, plants, animals, etc. The ground floor of the houses, with a public waterfront on one side and agricultural fields on the other, proved a perfect setting for the disurbanists and Wright to construct a kindergarten

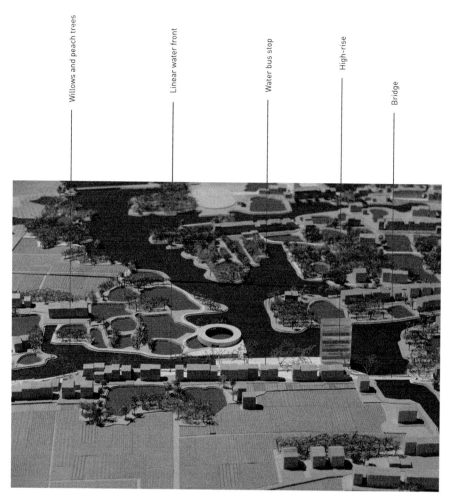

Willows and peach trees

Linear water front

Water bus stop

High-rise

Bridge

Fig. 5.18 Model photograph: the new bridge connecting the waterfront of the different *yu*. Model by author in collaboration with Lab-U, EPFL. Photograph by author

that teaches children about making things from an early age. Kindergartens and other facilities will be in a recurring pattern along the waterfront.

Collective vegetable gardens

As more migrants are expected to settle in the territory, a more complex and all-encompassing population structure will emerge, which calls for more diverse types of residential space besides today's peasant houses which include the owner's house and some spaces for low-income migrants. A large part of the population will live in apartments, studios, coliving spaces (including student accommodation), and other typologies without a private garden. Higher population density also generates new demand for vegetables. The introduction of collective vegetable gardens, which seems unnecessary in villages, could be relevant to the new conditions of the third space and introduce a quality that is unusual in today's urban areas, where there is no space for this.

The collective gardens will be constructed at the back of the peasant houses, where small orchards and water ponds are located usually. The gardens will be connected by small paths, which provide more private access to the houses at the back. In this way, today's disorganized vegetable fields which occupy all the available plots in the village will be replaced by a system of linear vegetable gardens close to the residents, a collective space shared by the farmers and others (Fig. 5.19).

Fig. 5.19 Model photograph: the collective vegetable gardens and other elements along the waterfront. Model by author in collaboration with Lab-U, EPFL. Photograph by author

Residential streets along the water

Residential streets, streets near shops, and other local facilities, where there are few cars and a mix of cyclists and pedestrians, are constructed in the third space. Again, it is one of the elements that is shared by the urban and the rural. The proximity to the water gives them a unique quality and makes this a model that can be repeated throughout the territory.

The residential streets along the water are continuous and form a lower network through the territory next to the "boulevards". They are intimate but also dynamic public spaces connecting an immense number of houses, facilities, and activities in the territory (Fig. 5.20).

Fig. 5.20 View: the impression of the residential streets along the water. Illustrated by the author

5.3.4 Industry and Energy Network

Industry in the third space has seen growth in spatial use. This growth takes the form of the dispersed micro-industries in every corner of the diffuse settlements and small-medium industrial platforms at the junction of county roads (Fig. 5.21). Today, the industry in the Yangtze River Delta is confronting serious challenges. The most serious constraint on industrial development in Tangqi, a typical town in the third space with a strong industrial tradition, is the space and the environmental capacity, especially with the increasingly strict policy measures to reduce energy consumption and pollution nationally (Ding 2012). Our imagined concept, taking into account the general tendency to upgrade and optimize the current industrial space, does not try to propose a comprehensive plan for future industrial activities, but simply explore one option: the industrial space as an energy producer.

The energy supply in the third space

The city of Huzhou's vision of energy[24] could be representative of the energy supply in the third space. Huzhou, a city with zero fossil fuel resources, relies heavily on imported energy: 73.7% comes from a local fossil fuel power station (44% from coal, 19.7% from oil, and 10% from natural gas), and 8.3% is directly imported electricity. The remaining 18% comes from local renewable energy, including hydropower, pumped-storage hydroelectricity, biofuel, wind power, and solar power. The latter is set to expand rapidly in the coming years. By 2020, Huzhou expects a total installed capacity of 7.95 mWh, with solar energy increasing from 3.3% to 16.4%. Wind power (1.5%) and biofuel (0.9%) do not yet supply enough energy.[25] Similar conditions apply in most parts of the third space.

The Yangtze River Delta benefits from many hours of sunlight (sub-tropical climate): Shanghai has more than 2000 h of sunlight a year, and some parts of the Delta reach 2200 h, which means, taking Shanghai as an example, 4700 mJ or 1307 kWh of solar energy per square meter (Zhang 2007).

The industrial space as a producer of energy

The 50 × 50 km study area of the third space has a diffuse distribution of small-medium industrial platforms bigger than 10 ha. In 2014, they occupied a surface of 12,883 ha. A rough calculation based on the photoelectric conversion efficiency of 15%,[26] a building density of 50%, and electricity consumption of 7000 kWh per

---

[24] The energy planning for a town like Tangqi is carried out on the city level, in this case Hangzhou, which is one of the three largest cities in the Delta. In contrast, the energy planning for the city of Huzhou, a small town neighboring Hangzhou in the north, is more representative of the third space.

[25] Source: The Thirteenth Five-Year (2016–2020) Plan for Energy Development of Huzhou, 2017, published by the Development and Reform Commission, City of Huzhou.

[26] The monocrystalline solar panel, the most technically sophisticated and the most efficient in photoelectric conversion, can reach an efficiency of 15% in massive production. The number of

Fig. 5.21 Existing industrial buildings and residential buildings. *Source* Google Earth. Illustrated by the author. Red: industrial buildings. White: residential buildings

capita demonstrates that the transition of all the industrial building surfaces including the industrial platforms to photovoltaic panels could fully support a population of 1.8 million, both for household and industrial electricity consumption. It means a population density of 720/km² could be self-sufficient in energy production, which is like the density of today's third space (in the 50 × 50 km study area, it is around 800/km²). If the population doubles in the future, the surplus demands could be compensated for by improving the photoelectric conversion efficiency, the densification of the industrial space, and the broader use of photovoltaic panels (e.g., on the roofs of the peasants' houses).

15% is representative of today's technological level and widely used in China (Zhang 2007; Wang 2010).

The local industry platform, a relatively intensive energy consumer, could benefit from the on-site production of energy with almost zero distance of transmission loss. A synergy of energy use could be imagined between the local industry platform and the dispersed peasant houses (which could also become energy producers by installing photovoltaic panels), profiting from their different consumption patterns (Fig. 5.22).

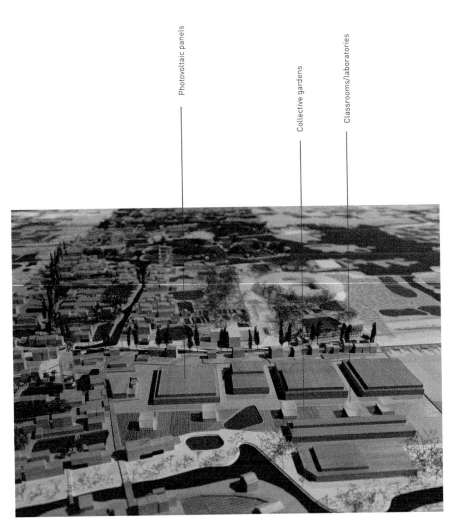

Fig. 5.22 Model photograph: the roofs in the industrial platform for energy production. Collective gardens penetrate the industrial platform, combined with classrooms and laboratories as part of the system of schools. Model by author in collaboration with Lab-U, EPFL. Photograph by author

Decentralized energy production

Throughout history, a balance was struck between the productivity of the arable land, the type of agricultural production, and the population density. Today, another balance between the radiation of sunlight, the distribution of industrial space, and the population density could be imagined. Far from a concrete proposal for an energy solution, the concept here serves to reveal the specificity and the unique potential of the third space.

The decentralized distribution of energy has been the presumption of the architects referred to in the *atlas of utopia*. Noticing the electricity giving the possibility to return to domestic work, Kropotkin sees the universal supply of electricity as a new motive for small industries, with an emancipatory power to free workers from concentrated industrial production, or workshops, all the way to their own houses (Kropotkin 1901: 154).

For Wright, the congested and vertically organized city has become "intolerable". The "electrical, mechanical and chemical invention" are part of the modern forces that create a condition not just for the industry as a sector but the whole society from living vertically to living horizontality, the forces "that are tearing the city down" (Wright 1932: 23–24).

However, despite the decentralization of the energy supply, the decentralization of energy production is not evident in the observation and vision of Kropotkin and Wright. The power plants as energy centers generate power distributed to the citizens; therefore, theirs is still a center-periphery relationship. Okhitovich criticizes Kropotkin whose idea still lays in the duality of the "center" and the "periphery" and proposes a radical organization of energy supply, namely one single diffuse network of energy production and consumption, in which both production and consumption are decentralized. With this network, the concentrated production and the diffuse micro-production do not just co-exist, but are united into one territorial whole (Okhitovich 1930).

In the view or Okhitovich, such a decentralized energy supply does not only support the small craftmen as Kopotkin imagines, but a support for a large scale and universal mode of production.

In the Maoist commune, decentralized energy production is evident, and many successors from those setups can still be seen today. The industrialization of the rural communes and the mechanization and electrification of agriculture are two of the main tasks of the national, socialist construction program.[27] During the Cultural Revolution, the people's communes were broadly constructed into the so-called Five Small Industries, including the small iron factory, small machinery factory,

---

[27] In 1958, Several Decisions on People's Commune were passed by the Central Committee of the Party, stating:

"From now, the tasks that our people are facing are: the rapid development of social productivity, promotion of industrialization of the country, the industrialization of the communes, the mechanization and the electrification of the agriculture through the social organization of people's communes, and based on the general direction of socialist construction proposed by the Party. Furthermore, gradually construct our country into a great socialist country with highly developed modern industry, modern agriculture, and modern science and culture".

small fertilizer factory, small concrete factory, and small energy factory (including small fossil fuel power stations and small hydraulic power stations). According to western scholars, it was the Chinese way of industrializing the countryside that not only leveled up productivity but also built a bottom-up, resilient system against natural catastrophes.[28] The construction of micro-hydropower plants was promoted by the guidelines for the electrification of the countryside: the codevelopment of the grid and the rural micro-hydropower plants (Shi and Zhang 2009: 523).[29] Qian, in his famous concept of the sixth industrial revolution—the one moving toward the knowledge-intensive agriculture industry—proposed to use agriculture with its territorial scale to fully utilize solar energy. This complete utilization, since photovoltaic technology was not yet available, included wind, hydropower, all types of plant products, biomass, biogas, etc. Agriculture is defined as a surface for receiving solar energy and thus should be implemented everywhere and transform the entire territory to an energy producer.

Imagining the industrial space as a new producer of energy follows the concept conceived by the people in the "atlas of utopia": a diffuse network with decentralized energy production. First, it attempts to reveal a unique opportunity to realize such an agenda provided by the specificity of the third space—in this case, a great quantity of decentralized, unused roofs of industrial platforms. Second, this concept attempts to add the well-known technical advantages of implementing such a network (also called Distributed Generation). These advantages include the possibility to collect multiple forms of renewable energy, reduce and decentralize pollution, reduce emissions, the capability of providing emergency power for public services, improvement of grid reliability, reduction of energy waste during transmission, etc.[30] Finally, it attempts to address the possibility of creating a more horizontal and equal condition in society, a condition that eliminates urban–rural dualism.

5.3.5 The Houses

Today, the villages own the rural land collectively, and each peasant family is allowed to occupy a small plot of land to build on. Recently, large five-floor buildings have replaced many of the three-floor dwellings built in the 1980s and 1990s. The ground floor of this new typology is often used for commerce and other activities, with a separate entrance. The first floor and the second floor are used by the owner's family, with external and "monumental" stairs to the main living room on the first floor. The upper floors, served by an additional independent staircase, are often rented out to

---

[28] In Guo's study, a comparison between the fluctuation in agricultural production in the 1930s and 1950s, clearly shows the resilience of stable productivity as a result of these operations, when climate, natural catastrophes, and technology are comparable (Guo 2013: 296).

[29] After the Economic Reform, the construction of micro-hydropower plants has continued. Till the end of 2014, 47,000 micro-hydropower plants produced ¼ of total hydropower in China.

[30] See Introduction to "Distributed Generation", Virginia Polytechnic Institute and State University, 2007. http://www.dg.history.vt.edu/index.html.

migrants attracted by the industrial estates developed nearby. There is no connection between the owner's house and the rest of the building. Behind the house, there are additional buildings with various functions: small factories, workshops, storage rooms, etc. There is usually a concrete surface forecourt, and high walls have begun to appear around these buildings. These were not a common feature in buildings from the 1980s and 1990s. It could be a sign of a new society in *desakota* with a more complex population, accumulation of private wealth, and a need for security.

Urbanization without new buildings

An incredible number of housing surfaces were built and later abandoned. They are used nowadays to store building waste. This is due partly to the generous compensation, in cash and/or residential surface (apartments or houses), which is awarded when those peasant houses are demolished to make room for the expansion of the city. But together with this demolition, the material, labor, and energy invested in the construction of the original settlement, the roads, and other infrastructure, and part of the ecosystem and landscape—including the *yu*, the fields, the trees, the fish-ponds—intensively maintained by the inhabitants, are also lost. The built surface of each house could easily accommodate double its current number of inhabitants, paving the way for an alternative model of urbanization, that consists of filling the existing spaces. One could say densification with hardly any new buildings.

A territorial housing program recycling existing peasant houses could be imagined. The newcomers no longer come to the city, but to the third space—the industrialized countryside.[31] Instead of the demolition-compensation model, the government could simply purchase part of the existing houses (the upper three floors, for example), transform them into new housing, and leave the lower two floors to the owner's family and his/her business. Since many peasant' houses are built in rows, the upper floors could be divided and/or grouped, which enables the co-existence of different typologies such as apartments, student studios, duplexes, triplex, and coliving spaces. A certain amount of social housing could be realized first to house the current migrants living in miserable conditions today. The different types of residential spaces could be accessed with existing and new means of movement. New corridors above the ground could be added to ensure passage and the necessary facilities.

The ground floors

Some of the ground floors will be sold and grouped for commerce, production facilities, and services. They could also be purchased by the public and transformed into facilities such as a kindergarten. Each house is allowed a front garden, no more than three meters deep without significant fencing. More transparency of the ground

---

[31] Although more study must be done, the population in the third space seems to be increasing quickly as of late. During the development of this thesis, the percentage of the migrants in some parts of the case study area in Tangqi has reached 50%.

floor, in terms of function, material, and space, is encouraged, to visually and spatially connect the waterfront and the agricultural fields.

Collective gardens

A collective garden is provided next to each building as a space between the building and the agricultural land. About 100 to 200 square meters of land for each unit is provided, which is enough to cover the domestic need for vegetables. In some parts of the "garden", there could also be a fishpond or an orchard, depending on the choice of the inhabitants. The collective gardens are also the back entrance to each building, connected by a path. The forecourts of the buildings are open and connected, especially when they are facing the water, as a public space in the village.

Tall buildings

Sporadic tall buildings—as in Broadacre City—are built. They are landmarks spread across the field. The ground floor of the tall building is a combination of a civic center, police and fire department offices, a water bus station, and a limited number of collective parking spaces that also serve other peasant houses. They are not "free-standing in small green-parks of their own" (Wright 1958: 133), but are close to the bridge in between the *yu*, as a continuation and conclusion of the row of peasant houses (Fig. 5.23). The tall buildings are carefully positioned to avoid shadow on the lower houses. The material of the tall buildings is light and transparent, which reflects the rich landscape around the buildings, and fades into the color of the sky. At night, they function as light towers in the dark and quiet landscape (Fig. 5.24).

Everybody has a view, everybody has a garden

The diversity of the space recalls Constant's New Babylon. The space could be articulated with terraces and gardens, to enjoy the fantastic view of the countryside from above. Mono-oriented typologies will be avoided, and a view is provided for each unit. Peasants receive the full right to buy and sell their properties. The housing agenda, a fundamental element that radically shows the rural–urban divide and a social division of labor, is detailed in the atlas and offers plenty of ideas. The disurbanist idea of the linear settlement is an organization of housing replacing both the city and the villages. The basic unit is a pod house—a simple elevated room, with openings in the front and the back, but not on the sides. In the competition entry for Magnitogorsk (Okhitovich, Ginzburg, 1930), this basic unit could be freestanding, or coupled, grouped, and linear by interconnecting the sidewalls; in the competition entry of the Green City for Moscow (Okhitovich, Ginzburg, 1930), countless numbers of units constitute an endless linear settlement crossing the entire territory, the sidewalls interconnected. There is no doubt that the essence of this linear settlement lies in its radical social agenda: its radical dispersion of living brought by the dispersion of production, the radical elimination of the urban and rural living typologies (urban houses, apartments, villas, etc.), even the elimination of the spaces

Fig. 5.23 Imagination: high-rise buildings as landmarks in relation to the water courses and surface. Illustrated by the author. Yellow: existing houses. Red: new high-rise buildings. Light gray: water surface

of men and women (laundries, kitchens, etc.). However, the essence of this linear settlement also lies in its universal living quality: every room should have cross venti- lation, open horizons (on both sides), spaciousness, and ideal natural lighting, and every room should have a view of the forest on one side, and the open agricultural field on the other side, without any obstacles. The unity of the urban and the rural is presented here not by equivalence in abstract indexes but by a specific spatial expe- rience. This universal living quality is also shared by Wright in his Broadacre City, despite the drastic difference in its ideology compared to the communist disurban- ists. "Air, sunlight, and land" or the "association with the sun, sky, and surrounding gardens" that the dispersed and horizontally organized Broadacre City provides are the fundamental elements in human life (Wright 1932: 45). Instead of disurbanists'

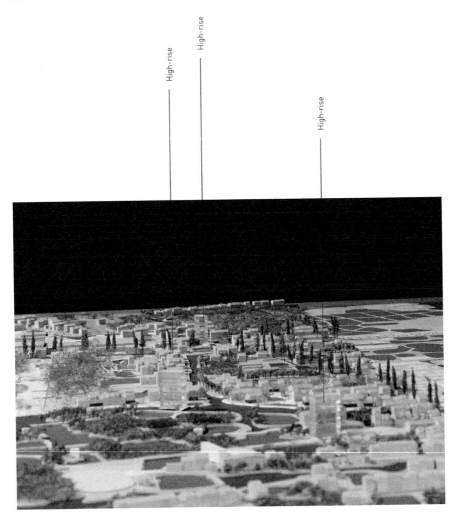

Fig. 5.24 Model photograph: high-rise buildings as landmarks. Model by author in collaboration with Lab-U, EPFL. Photograph by author

territorial forests and agricultural land, the view out of one's living unit—the house, is the one-acre garden for each house, especially the garden as a constantly repeated element in Broadacre City (Wright 1932: 60). The plan of each house has an open character to integrate the garden as part of the house, an external room.

During the construction of people's communes in the Maoist era, planning and architecture were significantly influenced by Soviet architects who held a different view than that of the disurbanists. The modernistic masterplan of Magnitogorsk was influenced by its former version designed by Ernst May. He was introduced as a replacement for the disurbanists after the competition of Magnitogorsk. This plan was imported to China along with other plans as an example of planning principles.

However, these principles are very critically implemented and complemented with agendas like those in the atlas. For example, the design of the house in the commune was also organized in lines, besides the principle of good ventilation and natural lighting. It dictates the productive space around the house (vegetable gardens, chicken coops, orchards, etc.) and suggests the "gardenization" of the residential area in general, to provide views on both sides of each house (Chen 1959: 40).

If this agenda of the elimination of the rural and urban could be, to a very limited extent, achieved by the spatial agenda. If every inhabitant had cross-ventilation and a naturally illuminated living space, views of vast natural and/or agricultural fields on both sides of his/her house, and access to a private or collective garden where his/her vegetables for own consumption could be produced. If this spatial agenda could be a step toward uniting the city and village into a new type of settlement, with a balanced dense population essential to create culture, and openness toward nature, a true city in the field, then the current organization of the settlement in the third space is a legitimate base for the implementation of such a spatial agenda. This base is composed of the linear organization of the houses in east–west rows, the linear public space along the water, the proximity of vast and varied landscapes, and the possibility to implement collective gardens…a spatial specificity that cannot be found in the current cities and remote villages.

References

Architecture Department, South China University of Technology (1959) Architecture planning and design of people's commune. South China University of Technology Printing Press, Guangzhou

De Witt N (1961) Education and modernization in the USSR. United States Government Printing Office, Washington

Engels F 1820–1895 (1942) The origin of the family, private property and the state. New York, International Publishers

Friedmann J, Douglass M (1978) Agropolitan development: Towards a new strategy for regional planning in Asia. In: Lo FC, Salih K (eds) Growth pole strategy and regional development policy. Pergamon Press. Proceedings of the seminar on industrialization strategies and growth pole approach to regional

Ginzburg M, Bard M (1930) (1930) Green City (Зереный горов. The Contemp Architect (современная Архитектура) 1–2:20–36

Guo Y (2013) 中國農業的不穩定性(1931–1991), 氣候、技術、制度. The Chinese University Press, Hong Kong

Kropotkin P (1901) Fields, factories and workshops: or industry combined with agriculture and brain work with manual work. G. P. Putnam's Sons, New York

Kropotkin P (1902) Mutual aid, a factor of evolution. Nature 67:196–197 (1903). https://doi.org/10.1038/067196a0

Langmead D, Johnson DL (2000) Architectural excursions: Frank Lloyd Wright, Holland and Europe. Greenwood Press, Westport, Conn.

Mao Z (1958) Sixty Points of Working Methodology (Draft) (工作方法六十條(草案)), In: The literature research center of the central committee of the communist party of China (eds.). (1992). The text of Mao Zedong since People's Republic of China (建國以來毛澤東文稿), vol VII. Beijing, Central Committee of the Communist Party of China Press, p 64

Marx K (1867) Das Kapital vol 1. Hamburg: O Meissner

Maumi C (2015) Broadacre City, la nouvelle frontière. Editions de la Villette, Paris

Miliutin NA (1930) Sotsgorod: the problem of building socialist cities. Mass, The MIT Press; 1st edition (February 15, 1975)

Okhitovich M (1930) The notes on settlement theories (Заметки по теории расселения). The Contemporary Architecture (современная Архитектура) 1930(1–2):7–15

Scalapino R, Yu G (1961) The Chinese anarchist movement. Berkeley, Center for Chinese Studies, University of California; Secchi B (1988) La Macchina Non Banale: Una Postfazione, in Urbanistica, n. 92

Shi Y, Zhang X (eds) (2009) Chinese academic canon in the 20th century. Fujian Education Press, Fuzhou

Snow E (1937) Red star over China. Victor Gollancz Ltd., Random House, London

St. Clair J (2004) Usonian Utopias. In: Cockbum A, St. Clair J (eds) Serpents in the garden: liaisons with culture and sex. AK Press, p 225

Viganò P (2012) The contemporary European urban project: Archipelago City, Diffuse City and Reverse City, in Crysler CG, Cairns S, Heynen H (eds) The SAGE Handbook of Architectural Theory. London, SAGE

Wang Y (2010) 节能减排:低碳经济的必由之路. Shandong Education Press, Jinan

Wright FL (1932) The disappearing city. W.F. Payson, New York

Wright F (1935) Broadacre city: a new community plan. Architect Rec, p 344

Wright FL (1958) The living city. New York, Horizon Press

Yiu A (2008) Atarashikimura: the intellectual and literary contexts of a Taishō Utopian Village. Japan Review, No. 20, pp 203–230

Zhang Y (2007) 长三角能源战略创新. [online] 中国网. Available at: http://www.china.com.cn/economic/zhuanti/cxsj/2007-06/06/content_8360655.htm. Accessed 26 Oct 2017

Chapter 6
Conclusion

Abstract The conclusion offers reflections on the *desakota* as a specific physical environment and its limit to the interaction between the exogenous and endogenous forces, resisting the latter while overwhelmed by the former. It is followed by a description of the "Elementary Metropolis" as a type of work. The "Elementary Metropolis" presents a research-by-design through a sequence of steps, a complete trajectory based on the elementalism through which the physicality of a metropolis can be explored through the research of its past and the imagination of its future (Fig. 6.1), a type of work that can offer openings to other research topics and discussions that are today emerging on the theme of development and urbanization across the urban–rural spectrum in Asia.

Keywords *desakota* · Type of work · Elementalism

6.1 A Reflection on *desakota*

A few reflections on the condition and the transformation of the third space, derived from my work, are presented here through an analysis of the *desakota* concept.

Similarly, the third space conditions can be detected today all over the world. The Horizontal Metropolis research demonstrates a broader research tradition on regions with dispersed conditions, with closely interlinked, copenetrating rural/urban realms, communication, transport, and economic systems, such as Città Diffusa in Northern Italy, *desakota* in Asia, the dispersed urbanization of the "Flemish diamond" in Flanders, Zwischenstadt in Germany, etc.

Among this body of research, the concept of *desakota* covers many territories in China including the Yangtze River Delta. McGee in his "The Extended Metropolis: Settlement Transition in Asia" (Ginsburg et al. 1991) defines three *desakota* types and tries to describe regions with an intense mixture of agriculture and nonagricultural activities. The Yangtze River Delta is described as a type of *desakota* that has experienced a rapid change in its economic situation (from agriculture to nonagriculture) in the past thirty years (the 1960–1990s). This change is quite different in countries like Japan and Korea where this transition is already further developed,

© The Author(s), under exclusive license to Springer Nature Switzerland AG 2023   203
Q. Zhang, *The Elemental Metropolis*, The Urban Book Series,
https://doi.org/10.1007/978-3-031-36409-9_6

Fig. 6.1 Imagination of a city in the field, the case of Tangqi. Model by author in collaboration with Lab-U, EPFL. Photo by author

and in areas like Kerala and Tamil Nadu (India) where the transition is slow. The development and urbanization of *desakota* have been interpreted by two knowledge clusters: firstly, the "above ground knowledge", which emphasizes the centrality of the cities, the role of "the extension of urban activities into rural areas", the role of institutional policy and globalization; and secondly, the "grounded knowledge", which gives "greater importance to the role of contexts, specificity, and local activities in the prevailing conditions of the urban transition". Although the majority of McGee's work is based on "grounded knowledge", he is well aware of the interaction of the two forces. From the 1980s to the 2010s, he further developed the model of *desakota*. In his first version, the *desakota* was discovered more around the major cities and the main mobility corridors in between those cities. Between *desakota* and the major cities, lay a "densely populated rural" area. Later, McGee observed more and more the increasing influence of the urban core expansion into the peri-urban and the *desakota* area, and the intervention of regulated "new towns" and "high-tech" industrial estates into the diffuse and in-situ urbanization. Consequently, the new version (McGee 2016) of the *desakota* model presents a noticeably more condensed relation between the urban cores and its peri-urban area and the *desakota* area, with the vast space of "incipient *desakota*" in between (Figs. 6.2 and 6.3).

The description by McGee is very powerful, and the model of the *desakota* has turned out to be a very useful cognitive tool through which the characteristics of urbanization in different parts of the territory and the relation between them are outlined. My work, which has been almost entirely a collection of "grounded knowledge", confirms this description to a large extent and shares the recognition of an increasing "exogenous" influence on the historically "endogenously" driven urbanization. Nevertheless, there are two reflections intended to complement the model.

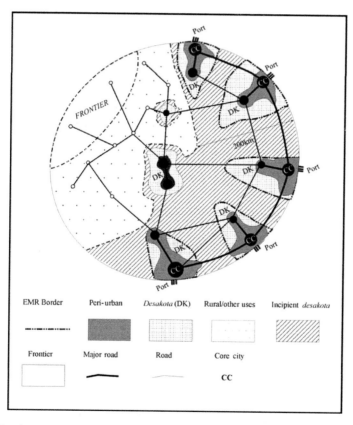

Fig. 6.2 Land use model of urban space at national level, 2007. (McGee 2007:70). The diagram shows a more condensed relation among the urban cores into its peri-urban area and the *desakota* area

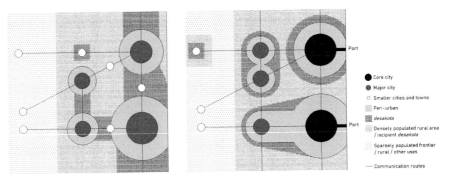

Fig. 6.3 Left: The interpretation of the *desakota* concept of McGee in 1991. The diagram shows that the *desakota* was discovered more around the major cities and along the main mobility corridors in between those cities. Elaborated by author based on the original diagram. Right: The interpretation of land use model of urban space at national level (Fig. 6.2) by McGee in 2007 (McGee 2007:70). The diagram shows a more condensed relation among the urban cores into its peri-urban area and the *desakota* area

The first reflection is that *desakota* as a specific physical environment[1] saturates the space between the "core cities" and the "frontier". There is no doubt that there has always been a certain dependency on the cities. However, this historically built territory started in the in-situ urbanization, both in the agricultural economy and the Maoist era, much earlier than the development of the cities took off. After decades of urbanization, it is no surprise that in many ways, a section of the *desakota* could be called a peri-urban area, or incipient *desakota*. The case of Tangqi presented in my thesis is a good example: it can be called a peri-urban area of the city of Hangzhou, as some parts get real estate investment from Hangzhou's inhabitants, and industries pushed out of Hangzhou are relocated to the small but regulated industrial platform in Tangqi. It can thus be called *desakota*, as in many parts, it maintains diffuse urbanization, the diffuse mixture of agricultural, and industrial production without any highly-regulated intervention. We could also call it an "incipient *desakota*" or even a rural area, as some parts consist of abandoned agricultural land, deserted villages, and large areas of hidden and beautiful nature with very few inhabitants. And last but not least, it could even be called an urban area, in the sense that its inhabitants are leading an almost complete urban life, in which only an extremely small amount of time and effort is dedicated to agriculture.

Moreover, the physical elements and their layers, presented in chapters three and four, form a strong continuity throughout the different "zones". Although, as McGee correcently pointed out, urban objects have been invading *desakota* and therefore transforming it into a more peri-urban condition, the water, the fishpond, the trees, the roads, the houses, the small industries, and the schools create an evenly distributed condition and a spatial character with a mixture of urban and rural, open and built, and artificial and natural, throughout the territory, making the boundary between

---

[1] Here, the avoidance of using the word "space" is deliberate, to make a distinction to "space" in the *desakota* model which is much more conceptual and broader in scale.

the peri-urban and *desakota* indefinable.[2] In comparison with economic, social, and technological advancement, the transformation of the physical environment is relatively slow. The materiality of the territory from past decades or even centuries is still visible. Other motivations and operations, such as the notion of maintaining agricultural production, the need to prevent floods and protect the ecosystem, the spontaneous construction of peasant houses, and the renovation of existing infrastructure, are stabilizing the organization of those elements. Therefore, a new question arises: when a *desakota* is peri-urbanized, or in general, urbanized, how could this process be supported by a physical environment with elements that were planned and constructed for much more rural circumstances? Or could we imagine a new type of urbanization within this unique landscape?

The second reflection is a hypothesis: there is a limit to the interaction between the exogenous and endogenous forces, resisting the latter while overwhelmed by the former; or, in Viganò's terms, between the "geometric rationale" and the "territorial rationale" (Viganò 2013). The study of the elements and their layers provides evidence for this hypothesis. A good example is the element of schools, which refers to the question of concentration versus decentralization. Exogenously, the concentration of schools since the end of the twentieth century, driven by the enhancement of efficiency in utilizing the teachers and facilities, has reached a limit in that a further step would not bring extra resources to improve educational quality. Endogenously, the distribution of the population, the commuting distance and safety of the students, the capacity of the infrastructure, and many other reasons strongly oppose a further degree of concentration.[3] The distribution of the schools, locally concentrated but globally still incredibly decentralized, is stabilized around this limit; one could even argue that the local concentration of the schools consolidates the decentralized system and prevents its performance from being overly inefficient. Therefore, the "conflicts" generated by this interaction "that are reflected in rural opposition to urban development, economic and social inequality, and environmental problems and that create difficulties of administration and governance problems" (McGee 2016) can be measured through the form of the space, by the degree of being centered/isotropic. Until now, despite areas like Kunshan as exceptional cases, the majority of the *desakota* have been influenced by the expansion of the urban cores and global capital in an incredibly isotropic way, which does not call for the creation of large-scale *tabula rasa* and construction of regulated colossal objects. For example, the electricity and information network have rapidly urbanized the lives of the rural inhabitants and barely modified the physicality of the space. Local factories have already connected as a decentralized production network, in which orders from all over the world are isotropically treated through all scales of productive spaces, even without passing through the urban cores. The urban "expansion" at the scale of the

---

[2] McGee also admits that "the zonal separation between peri-urban and "*desakota* regions" while experiencing an ongoing expansion of peri-urban and *desakota* boundaries should still be recognized". (McGee, 2016).

[3] Another example is the construction of New Socialist Villages that could be locally perceived as a concentration process but globally as a dispersion process.

desakota has not triggered a broad and fundamental reorganization of the form of the territory. In other words, the form of the *desakota* as an isotropically organized territory is capable of supporting a much more urban condition, or the *desakota* is regenerating itself into an isotropic city.

6.2   The Elemental Metropolis: A Type of Work

Urbanization in China, especially the most urbanized areas such as the Yangtze River Delta, has attracted a great number of scholars not only from China but from all over the world. This thesis draws attention to a specific part of the territory in the Yangtze River Delta, which has been overlooked by current research, policies, and design.

This part does not factor into the typical perception of China, as a country with highly developed metropoles such as Beijing and Shanghai, and undeveloped rural areas whose populations are migrating rapidly to the metropoles, leaving behind empty villages. This specific part of China referred to in this thesis has no metropolitan urban cores, a concentration of arterial infrastructures, and vast industrial estates and development zones. Nor has it remote and disconnected villages with a beautiful idyllic landscape. This particular area has diffused but also dense urban cores, an extremely well-developed water and road network, and a mixture of decentralized industries and high-productive agriculture. It is the third space. Vast but invisible, the third space is full of content but considered as being empty. Moreover, it is transforming fast but lacks planning. It is less the result of a rapid contemporary intervention stimulated by national policy and foreign investment, but more a consequence of a long history and tradition of endogenous and collective interventions on the territory.

My thesis, apart from a general description of the long history of diffuse urbanization, the clarification of the "third space" as the research subject, and a brief atlas of utopian works, has two main sections: a description of the elements of the territory, and an image of how that territory can be developed. This image is only an opening, not a finished picture: the elements are inexhaustible, and the image is never complete. Even though the description of some elements is shorter than others, more layers can still be found, and their choice is questioned. Although the vision of the systems could be criticized as premature, it is nevertheless important to present the full trajectory: from description to the image; from elements to layers and systems; from the local to the territorial scale.[4]

Firstly, the life of an element from its inception to its imagined future is presented as a continuity through this trajectory. In many cases, the life crisis of an element is twofold: it is embodied in its past but also its future. For example, it is inconceivable to allow an ever-increasing dispersion of industries or relocate all the dispersed industries into a handful of centrally organized industrial estates. Therefore, this

---

[4] The methodology of the work is strongly influenced by the work of Secchi and Viganò, especially "La città elementare" (Viganò, 1999). A further analysis of their works could be added to the thesis.

thesis attempts to go beyond the discovery of certain overlooked functional aspects of the current situation, as Kropotkin did with the decentralized agricultural and industrial production at the beginning of the twentieth century. It attempts to be more than a manifesto in which new interventions must be carried out radically detached from its predecessors, like the work of the disurbanists and Wright. In my work, I try to imagine a new phase of elements according to their own history and logic which still deliver a significant contribution to the design process.

Secondly, the elements are not meant to be treated individually or equally. The design and construction of different elements in the third space today are disconnected from one another. The representation of the layers intends to connect different elements on the territorial scale and construct a first interpretation of the structure of the territory. The systems attempt to shape certain forms in which the different elements perform collectively. Above the basic description of the elements, an upper level of structures, forms, and figures, starts to be extracted.

Thirdly, the vision of systems starting from the elements adds new content of perceivable and "everyday" qualities to the design. The depiction of the systems through the spatial elements makes it almost visible: you can picture your way to the school in the middle of the orchards or along the fishponds or waiting for your kids under the peach flowers on a plaza in front of the school. The abstract and even ideological choices between rural and urban, between concentration and decentralization, between artificial and natural, are dissolved and represented in the concrete choice of experience, or the quality of everyday life.

A type of research, especially the thematic mapping, layering, and visual build-up, follows a common protocol taken in the research of the Horizontal Metropolis, directed by Prof. Viganò in EPFL in Lausanne. This protocol enables the connection and the comparison of own research with a broader research field that meets similar conditions and territories all over the world. Together, common problems and risks, like the environmental crisis, mobility, climate change, and energy shortage, can be looked at. Broader themes can be discovered and studied such as the urban–rural divide, the mix of industry and agriculture, the compact or the dispersed, the concentration or the decentralization, the "living together", etc. Earlier visions and projects can be learned from and guide new ones, such as the Broadacre City by Wright, the Green City by the Russian disurbanists, and the Brussels 2040 Horizontal Metropolis by Secchi and Viganò.

Recently, in light of climate change and the on-going social transition, new research topics such as "sponge cities", the hydrological territories, urban metabolism, and political ecology are emerging in the context of development and urbanization across the urban–rural spectrum in Asia. The need for the understand of the physicality of a territory, in terms of its construction in the past, its current situation, and its potential future, has become evident and urgent. The type of this work is an attempt to construct such an understanding and offers openings toward the emerging topics and discussions. It allows continuous additions and revisions its content. The further and deeper description of future elements and layers not only enriches it as an achievement of "grounded knowledge" for future use by policy-makers and designers but also constantly generates new perceptions of and themes relating to the urbanization process.

References

Ginsburg N, Koppel B, McGee TG (eds) (1991) The extended metropolis: settlement transition in Asia. University of Hawaii Press, Honolulu

McGee TG (1991) The emergence of *desakota* regions in Asia: expanding a hypothesis, in Ginsburg N, Koppel B, McGee TG (eds) The extended metropolis: settlement transition in Asia University of Hawaii Press Honolulu

McGee TG, Lin G, Marton A, Wang M, Wu J (2007) China's urban space: development under market socialism. Routledge, New York

McGee TG (2016) Città Diffusa and Kotadesasasi comparing diffuse urbanization in Europe and Asia. In: A project in the making. International roundtable conference: territories of metropolis: compactness, dispersion, ecology: comparative perspectives between Asia and Europe, Shanghai, China

Viganò P (1999) La città elementare Milano Skira

Viganò P (2013) Urbanism and ecological rationality. Two parallel stories, in Pickett STA, Cadenasso ML, McGrath B (eds) Resilience in ecology and urban design, Springer, Heidelberg

Appendix A
Green City: Reconstruction of Moscow (1930)

Moisei Ginzburg, Mikhail Barsch

(Translated by Ekaterina Andrusenko, Qinyi Zhang)

When a person is sick, a doctor prescribes medicines. But it is more accurate and cheaper to prevent illness. The socialist type of medication is "prophylaxis". When a city is unhealthy, i.e., the town with all its attributes: noise, dust, lack of light, air, sun, etc.—turn to the medicine: the cottage, the resort,—the city of rest—the Green City (Fig. A.1). This is medicine. It is in the presence of the urbanization, it is necessary, and we just cannot dismiss this. But nothing could be ignored—this dual system of poison and antidote is a capitalist system of contradictions. City, however, must oppose the socialist "prophylaxis", and this method destructs the city with all its specific attributes of urbanism and creates such a way of settling humanity that would solve the problem of labor, recreation, and culture as a single continuous process of socialist life. In this case, people have to leave the place of their resettlement not for medicine (light, air, sun, and greens) but for the need of changing places, in the diversity of nature, overwhelmed by the thirst to see the new, expand our horizons, push the boundaries of our consciousness. Does the task of "holiday town" disappear altogether?.

However, it must become part of another broader challenge immediately. This task is the social reconstruction of the existing city and the functional unloading of a massive building cluster and urban volumes. So the "Green City" of Moscow must merge into the overall system of Moscow's reconstruction—it must become the first step in the number of measures for the "city reconstruction". Thus, first of all, it is necessary to establish the main ways of this reconstruction. In Moscow, we have today a population of over 2 million. Its growth continues, and if we consider it normal and do not take any measures to regulate it, what awaits us in 3–5 years? The population is over 3 million. The housing crisis leads to the fact that, despite all the norms and rules of building, all green plantations, which are still relatively abundant

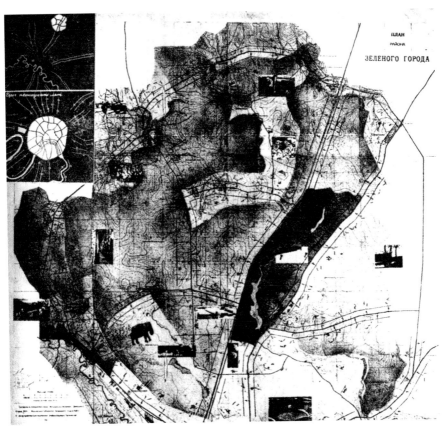

Fig. A.1 Competition entry of the Green City. *Source*: Ginzburg, M., Bard, M. (1930) Green City (Зереный горов). The Contemporary Architecture (Современная архитектура), 1930, 1–2, pp. 20–36

in separate possessions, are being destroyed. The first 2–3 thousand received from Avtostroi cars will turn all the streets of Moscow into a moving hell without exception. The number of accidents will become astronomical. Cars will not improve the traffic conditions—because they will be forced to move with the speed of the snail. Dust, tightness, noise, and commotion will give rise to such nerve diseases, which we do not even know by name. Our children's population-our shift will be doomed to degeneration. These horrors we portray are not exaggerated; it is also not a prediction. This problem seems a solved arithmetic equation with all known quantities, which is now just wished away.

However, by the end of the five-year plan, this task will have to be solved at all costs. This is the only way to address it: dynamite, which explodes entire blocks, expanding the artery is not decisive. The spent MILLIONS will only affect the palliative, time mara, little that radically changes:—"Green Cities" will have to devour other millions, also creating a palliative—to the solution—according to the path traversed by the old principles. We have the right to demand from Moscow more healthy and organic transformation, which can be called the ways of "socialist reconstruction". How do we imagine these ways?

Resistant and fixed systematic withdrawal from Moscow and resettlement on the USSR:

(a)   Moscow industries
(b)   Moscow scientific bodies and laboratories, universities, etc.
(c)   Administrative bodies,

which are not connected with coherent threads (local raw materials, etc.) with Moscow. Nowadays, this does not require any capital expenditure because the process is long and gradual. This is a clear trend, which must be persistently and planned to achieve—without making any deviations toward even small extensions of these institutions in Moscow. The funds allocated to these extensions must be transferred to the creation of the first points of the new location of these institutions at more appropriate places for this purpose. First and the basic measure will lead to a gradual decrease in the population of Moscow. The exact figure of it can be calculated. In any case, this will dramatically change the character of Moscow, its specific weight. Then another measure becomes possible, the resettlement of the remaining working population of Moscow not in Moscow itself, but along the highways connecting Moscow with other nearby centers. This maximally uniform and free resettlement of the proletarians, Moscow, and agricultural proletarians surrounding Moscow should be built on the principle of maximum human closeness to nature, the highest possible hygienic conditions of existence and the most perfect economic and cultural services for a person by collectivization, high technology, and industrialization. These principles are set out by us in the following chapters and are the basis for designing a Green City. Note that the architectural solution of settlement in a Green City is only of the possible technical variants, that the number of these options is probably very large, by the same principles. The more we unload Moscow, the freer this settlement will be. The harder it is, the denser the building tape will be. However, it will always be a line of free development, the density of which will be regulated by the possibilities of implementing it with the cheapest materials within a limited amount of time.

From the resettlement of Moscow industry, scientific, and administrative institutions, we could highlight following measures: a ban on the construction of any buildings in Moscow and systematic planting in empty areas. All construction in Moscow should be reduced mainly to the greening, with the ultimate goal of turning Moscow itself into a "Central Park" with cultural and recreational facilities. It is a point of intersection between all "strips" of the socialist settlement of Moscow. From an economic perspective, this is a completely painless process. We are forced to use the area of all existing buildings. BUT WE SHOULD NOT MAKE ANY NEW CAPITAL INVESTMENTS IN EXISTING MOSCOW. This "patient" is waiting for the natural wear and tear of the old buildings, and the execution of the amortization periods after which the destruction of these houses and quarters will be an UNHAPPY disinfection of Moscow. We will leave and carefully preserve the most characteristic pieces of old Moscow.

The Kremlin, pieces of nobility Moscow with the streets and mansions of the Arbat and Povarskaya, partly Prechistenka, pieces of merchant Zaryadye, Zamoskvorechye, Myasnitskaya trade, and proletarian Krasnaya Presnya. All the rest, we must persist in turning into a grandiose park, in which there will be freedoms; lots of remaining administrative institutions, scientific centers, and universities are serving only the population of Moscow, audiences, stadiums, water stations, zoos, botanical gardens, flower gardens, nurseries, and hotels. In other words, this is the CHEAPEST METHOD FOR RECONSTRUCTION OF MOSCOW. Despite this radical approach, it is the most real and economical way to solve the Moscow problem. IT IS DRASTIC. It destroys the evil of the big city. It simultaneously solves the problem of eliminating the contradictions between the major center and the province, between the city and the village. It organically solves the problem of "Green Cities". And finally, it treats the needs of the upcoming generations, allowing them to painlessly build their homes according to their requirements, which will be immeasurably higher than ours.

Moscow is gradual and depending on the possibilities requires NOW only the imposition of the TABU on new construction and the so-called expansion. Secondly: today's working construction of the building is thought of as the cheapest construction from local materials, in fact, the question of standardization and industrialization of construction by our specific capabilities. High-rise construction technology and high quality of housing here must be combined with the use of our cheapest materials and give the maximum effect with minimal economic costs. We know that our project of Moscow's socialist reconstruction will evoke the screams of homeless people, restorers, and eclectics of all stripes, but we are firmly convinced that these radical proposals are the only real and feasible plan, entirely possible economically today and inevitable tomorrow.

The primary organizing principles of the socialist "Green City".

We consider the main principle of socialist organization to be the fuller socialization of all economic, production, and service processes of the given collective, including nutrition, education of children, study, washing, and repair of linen, all types of supplies, etc. in these concepts.

The socialization of all these processes, following the example of the socialization of any other production processes, undoubtedly has advantages over any individual economy. It allows carrying out these processes with the highest level of technology, using modern machines, transport, and skilled technical and labor. Thus, the rationally conducted collectivization of these processes leads not only to a reduction in the cost of production but also to an improvement in the quality of the products produced. In the future, we will show how this fundamental principle is refracted in particular issues of nutrition, child upbringing, education, cultural services, etc. However, now it is necessary to note the unique power of the socialized economy, consisting in the possibility of active management and planning by the whole economy. It should be noted that with modern technical capabilities (auto, telephone, radio, television, airplane)—this concentration of organization and planning—all possible at any size of this economy, that it not only allows but also calls to life the need for the decentralization of individual elements of this economy.

In the future, we return to this idea. However, even here we point out that we see a fundamental difference in matters of production and consumption and that the mechanical extension of the principles of production to the region and consumption would be extremely erroneous and harmful.

Production is always the same, standard, and impersonal. Consumption encounters some individual characteristics of the consumer. Homogenous production is almost always rationally centralized. Consumption is almost always in the interests of the consumer rationally decentralized. Each factory must produce one product, one standard, in one place, with one streaming system. But this does not mean that the consumer should use only this product: he can consume it in different areas, at different hours, in different conditions. The mechanical transfer of the principles of production to the principles of consumption would inevitably follow the path of infringement of the interests of the consumer.

Similarly, we believe that the task of the fundamental importance of the socialist organization of society is the creation of conditions most favorable to the development and the flowering of each person. The socialization of the production and economic processes of the collective is valuable, which allows all the resources of the collective to use for this purpose, to create real prerequisites for maximizing the development of all the creative requirements laid down in each human individual. We must give ourselves a clear view of the fact that by combating the capitalist organization of society and the bourgeois individualistic way of life, we are fighting the methods of individual production (economy), where the old family way of life, built on the principle of private property and the exploitation of one member family to others. It is our real enemy, and this requires a brutal struggle.

But this cannot be used to make erroneous conclusions (often, unfortunately), that it is necessary to destroy the individual characteristics of the individual, to shave all under one comb: this means, together with the water, to throw out the baby himself from the bath. Because we are waging a struggle against the capitalist organization of society and with the bourgeois family and everyday life precisely because these latter destroy the human personality of the majority, turn it into a standard machine, providing hypertrophied opportunities for individual development to a few creams of the ruling class.

And, finally, a fundamental principle of the socialist organization of the collective, we believe, is an organization built on production. Concrete production is the real prerequisite on which the organization of the mass-production building, its development, culture, scientific thought, study, etc. These are industrial enterprises that at different times cannot yet be scattered far from Moscow, governmental institutions, and cultural facilities, which, due to Moscow's importance as an all-Union center for planning and administrative work, will retain their significance for a long time. They will either be scattered in the central park of culture and recreation or will be removed from Moscow along with the tapes of its resettlement. But, leaving aside the indigenous production of the inhabitants of the city, one cannot ignore the industry that should appear on the territory 3 H for the construction and maintenance of this first link of the future Moscow.

The local industry of the Green City and the cultural organization of its territory.

The first task of this industry is the task of constructing all those buildings that ensure the organization of the city and, in the first place, the creation of housing stock. The 3 G is the housing and construction industry. To clarify the tasks of this branch of the industry, it is first of all necessary to find out what the product will be manufactured by it. Three cardinal prerequisites determine its main features. The general economic and financial scheme for the development of the socialist construction of the USSR, in the sense that it is more profitable for us now to spend less capital with fewer amortization terms, in order to even completely renew the entire housing stock in 15 to 25 years, rather than spending large sums with a depreciation period of 70–100 years.

The extremely rapid growth of our nutritional needs and economic opportunities, will make in 10 years our best constructions worthless and reduce its cost at least twice. Approximate cost of a cubic meter of such construction, including all equipment, is about 10 to 12 rubles. The idea of Green City has economic benefits. Every year large-scale forestry gives a certain amount of construction timber for the frame and waste for fibrolite production. However, they will be always less wood than it needed for the construction industry. Undoubtedly, we will have to use the imported forest, delivered from the North along the Yaroslavl railroad, etc.

Area of Sofrino, lying almost on the border of the "Green City", supposed to become a center for construction industry, which includes:

(1) plant for the manufacture of wooden parts of structures
(2) assembly lines
(3) furniture factory.

For this, we need to switch construction method from small manufactories to large-scale industrialized production.

These three prerequisites determined the nature of our construction industry, manufacturing-factory-made parts with minimum weight and mounted on site with a light crane in the shortest period. Wooden prefabricated frame from standard bars, fibrolite shields, fiberboard or wood frame window frames, and flat hot cement or roofing on a wooden base are the main constructive features of this construction. Not at all lowering the quality of this housing—they allow it to be carried out today by $100°$, in a factory way, they can be installed on site in a few hours and, most importantly, allow.

(1) factory for the manufacture of wooden parts of structures
(2) assembly
(3) furniture factory.

Parts of sanitary and specialized equipment necessary for complete installation of the house are brought from the places of their production and stored in individual warehouses. Woodcutter area in "Green City", directly connected with the construction industry, will be very big (about 10,000 ha). Using existing sawmills, we assume that this woodcutter area will be located south-west of the station "Bratovshchina". This site is conveniently connected by railways with the housing and construction industry.

Here will be located objects, close to Kurovo village:

(1) central laundry
(2) repair shops.

An incredibly complex problem is the organization of the local food industry, which will be composed partly of imported raw materials, in part from the use of local agriculture resources. First one is concentrated on the southern border of the city.

The water of river Uchi (close to the station "Pushkino") will be used for industrial centers. "Pushkino" station serves as well a significant "Hkchad farm" of the "Green City", including warehouses, warehouses of food products (salt, groats, sugar, etc.). Moreover, it serves warehouses of household and cultural services, etc.

Behind storage facilities there is a food factory consisting of:

(1) bakery
(2) candy production
(3) flour storage
(4) refrigerator
(5) meat, fish, game production plant
(6) sausage production
(7) canning production.

The rest of the food industry is built on local products produced by gardening, dairy, and poultry collective farms organized on the territory of the city.

The area of the village of Cernskoye-Nagornoe-Pervomaiskoye, as the most tree-less, is supposed to be arranged for farming. District Petushki-Daryino-Matyushino-Zimogory is supposed to be organized berry orchards. At the junction of truck farming and berry fruit in the Zimogor district, a canning plant is planned to process the products of these farms. To the truck farm, for the convenience of acceptable manure use, the area adjacent to the north of the Tzernsky up to Mitropolia, adjacent to the organization of a dairy farm with the following production facilities, adjoins the north:

(1) the central farmstead of dairy cattle
(2) the butter mills
(3) the cheese factory.

The Eldigino-Semenovskoye area is supposed to be organized for poultry farming. Besides, the Suhodol-Chapchikovo region is favorable for the organization of dairy farming. This area includes the general food service network of "Green City". Fewer territories of "Green City" are allocated for the organization of the housing construction industry, food industry, collective farms, and forestry. Within the limits of "Green

City", there are still substantial green massifs that need to be maximized for the cultural impact. These arrays are divided into groups:

(1) parks of culture and recreation
(2) a zoo
(3) a botanical park.

Park of culture and recreation

(Park of culture and recreation) stretches a wide strip between the Northern Railway, etc. and Yaroslavskoe highway from the village of Bratovshchina to the Spasskoye platform and somewhat north, including the Skalbu River and the possible organization of two large lakes here. From the south and north of the park, there are two large sports centers with stadiums, two large auditoriums, with adjoining rest areas, canteens, and buffets. The administrative buildings of the city are located on the territory of the park:

(1) Green and City Council
(2) Administration
(3) Transport and Road Management
(4) Police and Fire Protection
(5) Supply and Food Management
(6) State and professions organizations.

Similarly, the park of culture is the central base of social education and the cultural supply. In the park, there are many exhibitions of various consumer goods. Any resident of "Green City" can order by phone any required item from this exhibition.

Zoo is located in the southwestern part of "Green City" along the Serebryanka River. The zoo includes not only a collection of different animals but also research institutions and animal nurseries. In the southeastern part of the Kostino area along the Kalba River, there is a botanical park, which is organized very similarly to the zoo. This institution specializes in exemplary floriculture.

Principles of resettlement and roads

If the basis of socialist resettlement is production, then the primary factor in its implementation is transport, including in this concept road economy and means of transportation. Taking into account the high cost of road maintenance and its operation, we were tasked with maximizing its use. The usual system of resettlement is built on a dual system of road economy: roads connecting individual centers to each other and internal highways or streets along which the construction of populated areas is carried out. We adopted the principle of settling the policy of using intercity or intercenter roads, as roads along which the settlement is simultaneously carried out. Thanks to this principle, considerable savings and simplification not only of a pure road economy but also of all sanitary maintenance networks and reduction of the total number of freight flows are achieved.

It is worth making the most approximate calculation of the length of the roadways in the typical urban planning and comparing it with the proposed "linear" method of settlement to see the specific economic advantages of linear settlement. But the linear principle of resettlement has other, hugely significant advantages. The main one is hygienic.

With the usual urban resettlement, urban blocks are formed; that is, they are nothing more than closed or semi-closed airbags of greater or lesser magnitude, depriving residents of these quarters of through ventilation, open horizons, spaciousness, and ideal illumination. We are striving for such a settlement system, in which all these shortcomings would be eliminated. But having accommodation by lines along roads, it is necessary to protect it from traffic noise and dust. Therefore, we decided on the following scheme: from the line of the road, retreating a minimum of 200–250 m of a continuous park strip, there are ribbons of residential buildings. Thus, each link of the housing has in front of itself 28–29 one side of the park strip at 400–500 m of width, and on the other, the vast expanses of green massifs of forests and fields.

It is not difficult to understand what exceptional advantages of hygienics give this principle of settlement to the inhabitants. But at the same time, this outstanding free settlement system seems to exclude the possibility of settling large masses of people. Namely, we encounter this problem in "Green City". This problem is extremely relevant for any reconstruction of existing large cities.

To solve this problem without prejudice to the inhabitants of "Green City", we resorted to the principle of the continuous construction of dwellings along the roads. And if you understand the conditions of life in this constant "building stripe", it is easy to understand that it is essential to have two-sided apartments. And this has to be taken into account when making interior or floor plan.

Absolutely no significance for the living is not already the fact that behind its closest neighbors, there are also living cells. The person, living in one of the units, does not care about any open space or amount of living sells close to him. Thus, it is evident that if it is essential for a tenant not to have any structures in front of his eyes on both sides along the lines of settlement, it is entirely irrelevant whether this settlement is continuous or consists of separate buildings.

Only two problems arise with continuous tape building: the task of fire safety and the issue of the possibility of spatial communication of territories lying on both sides of the settlement. The first of these tasks is resolved by the device from time to time firewall walls or by the periodic introduction of links from non-combustible materials. The second—the fact that the entire residential floor is raised to a height of 2.25 m on individual wooden supports, which, with a frame structure and, if necessary, to rise all the same housing from the ground is rational and economical. Thus, there is also a need to have a canopy over the entrance (the entrance is under the ceiling) and the presence of large comfortable terraces under the house, entirely used in the summer. Also, the presence of this floor on individual supports also resolves the issues of private communication.

Descending down the stairs from the living cell, the occupant enters the covered space, where they can get to where it is needed. This covered space protects against rain and snow. Do you need to make these periods also warmed? It is a fundamental issue, to which we respond negatively. The economic disadvantage of the isolation is quite apparent both regarding construction costs and concerning exploitation costs. Besides, these warmed transitions or corridors in houses of a consistently socialist type with clearly differentiated functions inevitably expand into whole streets. Do I have to insulate these streets? On the one hand, correct physical education and, on the other, the proper organization of all household processes can make redundant warm transitions. And this is all the more desirable because the insulated communication elements are also sources of noise and anxiety; it is precisely the well-known short-coming of corridors of ordinary hotels that deprive living cells of the possibilities of absolute peace and independence.

Speaking about the principle of linear settlement, it is necessary to indicate that an abstractly taken straight line practically should acquire an unforeseen figure. Having as a goal to bring a person closer to nature, she must fully follow this "line". It should enable the person to use all the uniqueness and richness of natural resources. If there will be a hill, our residential "line" will rise on it. Close to the lake "line" will bend around it. Falling into the forest "line" can scatter by separate links between trees. And, finally, it is necessary to point out the important fact that when entering the district, it does not change its principle, it stayed the "line".

The inhabited band of a settlement stretched along existing and well-maintained roads carries out a smooth and accurate distribution of Green City's population. Railways in the north are functioned as the primary delivery force. There is a station approximately every 2 km. We provide a comfortable transfer of passengers from the train to the car. However, along with the railways, we also have Yaroslavl highway, turned into an exemplary motorway, which allows to develop the maximum speed of a car and to deliver a passenger to your accommodation without interruption.

Bus station

There are bus stations along the highways. An auto station is located at a distance of approx. 650 m. Furthermore, the bus station in the direction of the residential area of "Green City" should not disturb the residents with noise and dust. However, it should not be too far. Thus, from the bus station to its habitation, the pedestrian crossing does not take more than 5 and 10 min of time. Near the bus stations are located and small ravines for taxis and own cars. The central office of bus station is located in the administrative center of "Green City".

The task of rest and restoring the labor: it is well known that air, sun, nature, and cleanliness are the first conditions for the best rest and restoration of forces. In this project, an attempt is made to give these conditions at the maximum limit. The living area of the cell has an area of 12 ml, a total cubic capacity of 54 slices of meat (including a staircase, a restroom, and showers), i.e., much larger than that in a conventional building assigned to a person. We consider this cubature, however, minimal not only from elementary hygiene but also from the relative spatial self-perception of a person. We believe that this need for spatial space is so significant that the whole solution of a living cell is built on it. The cell has two-way lighting and through ventilation. Only two-way illumination gives some completeness of space and perception of nature. Sunrise and sunset, nature on one side and the other—all this is not a luxury, but a requirement. The windows are all the width of the walls, from floor to ceiling. The sun permeates through the living cell. The windows are folded in harmony, and the residing cell turns into a covered terrace, surrounded by greenery. The room almost loses its "room" specificity; it dissolves in nature, becoming in it only a horizontal platform. This function of the room, a function that maximally connects the dwelling with nature, is understandable to us so far only in the summer, certainly, under traditional physical education, it will be necessary at times and in winter. The regime of winter TB sanatoriums convinces us of this. Linen awnings drop in the summer above the windows, giving a shadow, and the so-called thermal blinds (wooden slats, sewn on dense fabric, felt and moving in vertical guides) in winter allow to regulate the degree of cooling of glass walls. The whole organization of the resettlement is such that before the windows on both sides of the expanse of greenery of gardens, parks, collective farms, etc. In each living cell, there is a wash basin, with each cell an adjoining shower room and lavatory. If the desire for hygiene and nature is mastering everything more and more by all, it is still not enough that we pay attention to the need to create the conditions necessary for the infinite deployment of a human personality, which most clearly characterizes a socialist society.

These conditions are required:

(1) For the possibilities of complete rest and quick recovery of forces.
(2) For the options of maximizing the creative strengths of man.
(3) For the opportunity not to exploit and not to be exploited. When they talk about exploitation, they understand it usually in a material sense.

Meanwhile, for capitalist society, the fact is also typical. Every minute moral exploitation of one person by another. Not to mention that the owner economically exploits the worker, and the mistress exploits the servant, the husband exploits the wife, the parents exploit the children or another way around, the fact of exploitation is in every detail, that the wife chooses her friends, considering the husband's opinion, the husband organizes his leisure, considering the opinion of his wife, and so on. It is only a moment to ponder this circumstance. To understand how capitalist society created the whole bitter way of life—at the one-minute exploitation of one person by another. The architect-builder of the socialist attitude must take this into account. And it is in the statement of this fact that each living cell of this project has a separate move and may not communicate with its neighbors. Let the husband and wife live side by side in two adjacent cells. Between them is a door that they can talk to. But the presence of a separate move in each of the two cells ensures that they can and do not communicate if they do not want to. The connection between two people, husband and wife, in particular, is a voluntary bond. At the moment when this connection becomes forced, in any of the domestic details—it becomes the fact of exploitation of one person by another. In the construction of living cells of this project, we tried to contribute to the complete freedom of a member of a socialist society and, first and foremost, to the total emancipation of a woman.

Food stations, recreations, and sports centers

The most complicated production task is to organize the nutrition of the Green City's population. This task is carried out by the combine (meat enterprises, refrigerator, sausage, canning, confectionery, bakery, and food stores) and collective farm (cheese, dairy, canning, and berry fruit orchards). All these enterprises produce semi-finished products. These semi-finished products are mounted in the special menus directly in the preparatory kitchens and canteens.

At each bus station, on the way to the residential belt, there is a dining room, each of which serves 269 people in 5–10 min walk from home. At the dining room, there are all the necessary kitchen facilities, where special trucks poultry processing plant distribute semi-finished products. Thus, the loading of dining rooms with semi-finished products is carried out extremely quickly by the main highways. The number of 250 people serviced by a single table, we consider the limit, in the interests of convenient and peaceful use of them. The "number" should be checked more quickly in the direction of reducing the number, so we developed another dining room for 100 people. A closer examination of this issue should clarify this figure. From the dining room are directly connected recreation rooms, the collective (large hall, terrace), and individual (separate booths for reading, games, talks, meetings, etc.). Recreation facilities for their part are directly connected with a small sports center, where its recreation area, showers, locker rooms, and storage of sports equipment.

Kindergartens, education

On the opposite side of the road, the residential strip is fringed with greenery. In these arrays are urban life, upbringing, and education of children, as well as the cultural life of adults. The closest to the house is the nursery. Each soil in 15–20 children occupies a separate structure. In the interest of preventing infection and more residential tape in the forest or the valley is a hostel for school-age children. They sleep in shared dormitories and spend all their time with their peers. There and then not far away is their school. Schools are also decentralized. They are only 75 in "Green City", and each of them is designed for just 150 pupils.

For particular care of children, several such buildings are connected by a general transition and are located in proximity to the adult dwelling, rhythmically in residential tapes. Mothers can feed babies and bed for their lives, spending a minimum of time. Almost in the same conditions, there are kindergartens. But the growth of children in kindergartens is already associated with the general life of adults. Children 4 to 5 years digging beds, taking care of the bird, get acquainted with the animals of the zoo and the vegetation of botanical fat. They are taught from an early age to three essential things: independence, mobility, and production processes. However, bigger schools have the advantages of being able to perfect equipment. And this highly technical equipment of schools is especially crucial for us precisely because in today's schools, we do not think of as a class for lectures by teachers, but as laboratories and workshops for independent schoolchildren under the guidance of teachers. To be able to have high technical equipment in our small schools, we come to the need to specialize each school in a specific category of knowledge: schools of science, agriculture, forestry, social sciences, industrial, housing, etc. Nevertheless, it seems to us that a narrow specialization in the middle school age would be dangerous. Technical secondary education is necessary. It is determined by the school cycle, which makes it compulsory for each student to receive periodic training in several schools, which together form a "complete cycle" of education. With an exceptional scope of communication and connectivity of all institutions with a motorways, this frequency of visits to various schools can be resolved in any form and any terms which is a matter of pedagogical nature. That is, an element of linking studies with a specific production area is brought in and the component of the movement—we consider it extremely valuable since it is they who expand the horizons and horizons of the growing child. Americans in their best schools teach geography from an airplane. While this training manual is not yet available for us, the bus will carry out some of its functions and make the entire school district a teenager's school.

Every school has one large lecture hall, workshop rooms, and laboratories. This lecture hall, workshop rooms, and laboratories are not only for the kids. They are for adults too. Educating kids about production from childhood is essential. We believe that many parents and adults will spend their leisure time in the workshops and laboratories of the school. There will be lectures, film shows, and concerts in the lecture hall. The school site should become a core of everyday cultural life for adults and children. On the territory of "Green City", there are also more specialized educational institutions. They are called "technical schools". Children who graduated from primary school can learn here about local production and craftsmanship. In the area of an agricultural technical school, there are faculties of,

(1) food processing
(2) agri-farm diary
(3) poultry processing.

On the territory of the housing construction industry: a housing and construction college with faculties:

(1) housing
(2) furniture and fittings.

On the territory of forestry, there will be forest technical school. Also, in the park of culture, we have two large auditoriums and cultural institutions. The central core of social education is located in the same plaza. This area is accommodating the whole cultural life of "Green City".

Appendix B
Notes on the Theory of Resettlement (1930)

Moisei Ginzburg, Mikhail Barsch

(Translated by Ekaterina Andrusenko, Qinyi Zhang. Published as Okhitovich, M.A. (2022). The New Settlement Pattern (1930). In: Barcelloni Corte, M., Viganò, P. (eds) The Horizontal Metropolis. Springer, Cham. https://doi.org/10. 1007/978-3-030-56398-1_10

A small introduction

The reconstruction of industrial and agricultural production in the U.S.S.R. sharply challenged all principles of building a dwelling as well as the social, geographical, and productive approaches related to it. The most modest and simple statement on all of these issues immediately and painfully strikes the attention and tense nerves of huge masses. Ideas in conjunction with the masses reached an unprecedented scope. In the heated brawl of discussions, each participant matured rapidly, receiving an infinitely fruitful experience, developing attitudes toward competing proposals and clashing ideas. In the fire of ideological battles, formulations were sharpened, positions developed, new ones were created, old ones were abandoned … The timely practice with unexpected force felt the meaning of the seemingly abstract theory, and concrete proposals began to be tested by the cold mind of the abstraction of historical judgments. Theoretical work rose from work offices, from the rustle of books, and became actual practice. Much of the past work became relevant for the future. The future, which was painted by some before, amazed the spectator of the era—this future turned out to be buried in yesterday. This alliance of theory with practice is never to be abandoned. It is time to sum up the first results. It is time to prepare your own, revolutionary theory of resettlement. For Lenin said: "WITHOUT A REVOLUTIONARY THEORY THERE CAN BE NO REVOLUTIONARY MOVEMENT". The proposed "notes" aim to provide an outline for a serious development of the Marxist theory of the resettlement of people and the geographical distribution of productive forces in their interdependence. In our notes, we deliberately eliminate

Q. Zhang, *The Elemental Metropolis*, The Urban Book Series, https://doi.org/10.1007/978-3-031-36409-9

the signs of the scientific apparatus as well as all elements of polemics. We give only a logical course of ideas, urging the reader to criticize them most severely.

The dwelling is always aligned with the place of production. A person always lives where he/she works. More precisely: a person ENDEAVORS to live where he/she works. The place of residence for a person is always their place of work, because labor turns an animal into a human. Labor is the only source of human development, or the only source of social existence, the only source of the HUMAN.

If the production is moving, then the place of habitation changes too—AFTER the production, FOLLOWING the production, TOGETHER WITH the change of the place of production; and the location of the dwelling of a human also changes. This is usual for nomads.

If a permanent place of work has been established, then a house is immediately built next to it. This is usual for the settled farmers. If a land plot is lost, the dwelling is lost too in search for a new place, and if a new land plot is found, a new dwelling is established, and the old one is abandoned.

If the place of work is abandoned, so would be the dwelling. Neither your poverty nor your avarice can stop this law of placing the dwelling near production.

It is not enough to say "near production".

The dwelling seeks to be as close to production as possible. The dwelling seeks to be located not even near nor around, but directly in production. It seeks to fit as close to production as possible, but of course in such a way as not to harm production itself but instead to supplement production.

For example, the dwelling of the peasant is not located on the plowed land. Here, the production surrounds the dwelling.

Both peasant's and craftsman's houses are places of habitation and production at the same time. Neither the peasant nor the craftsman go or ride to work.

The situation is more complicated in the case of manufacturing and industrial capitalist production. Of course, the Putilov worker does not live in Pskov. Of course, he will not live on the outskirts of Leningrad, nor in the center of the city. Even if a grand mansion is prepared for him in the center, he will always strive to live as close to the place of production as possible.

The need to use transport to come to work is a punishment for him. This is a misfortune for him, as an inevitable result of the crowding of people around production sites. Housing which is far from work is not proper housing.

A peasant who has received an allotment far from his village will inevitably move to another; if this allotment lies within the possession of the village, but far beyond other land plots, then the peasant, having obtained it, will settle there. If this one is an independent possession, located far from all the settlements, the peasant will simply start his own.

If the area of his work is vast, he will be torn apart. To keep up everywhere, in addition to his main, basic home in the place of his basic work, that is, the work that takes away most of his time, he also builds special "zaimki" (land plots) for temporary, but sufficient, long-lasting other occupations, the movement to which and movement from which back to the dwelling would take him the time necessary for the main production.

Thus, for example, "ZAIMKI" are made for forest, fur-bearing, hunting, fish, berry, mushroom, and other activities with the main occupation of agriculture.

The same is true, although for quite different reasons (which means—not the same thing) for the modern American farmer. The car, which is available to him, is not fast enough to deliver him to the production site in time. Here, the role of the car is different—the car should (like a telephone) connect the various processes occurring simultaneously in different parts of the production territory, and it must turn individual private production efforts into one common product.

Thus, homelessness does not only mean the absence of the space that belongs to me. HOMELESSNESS also MEANS THAT I DO NOT HAVE AN AREA DIRECTLY IN PRODUCTION. According to objective reasons, independent of the will, the person settles in the place.

On the one hand, the forms of production strictly determine the remoteness of people from each other, the compulsory remoteness of people, up to the fact that people live in the wild, abandoned, idiotic rural life. On the other hand, they also create the proximity of people, the forced closeness of people, hence the tremendous calamities of CITY CROWDING.

The desire to eliminate homelessness (for what kind of house this is, when it is far from work) under certain circumstances leads to crowding. The desire to eliminate crowding under the same circumstances inevitably leads to homelessness. The reasons for isolation, the causes of crowding, the conditions for their occurrence, and therefore the conditions for their possible elimination must be sought in METHODS OF PRODUCTION ALLOCATION.

The moment of opposition between the city and the countryside is the most interesting and decisive in our thoughts about the sources of housing isolation, abandonment on the one hand and causes for housing crowding and concentration on the other.

The city is the result of a social division of labor as well as the result of the separation of craft from agriculture, or otherwise the separation of the PROCESSES OF PROCESSING OF THE PRODUCT FROM THE PROCESSES OF ITS EXTRACTION.

The city, therefore, is not a product of some physical-geographical, i.e., natural conditions, but a consequence of PUBLIC, SOCIAL CHANGES. Raw materials are available everywhere, both as the product of nature and the product of agriculture. However, the city, industry or, more accurately, the manufacturing industry does not appear everywhere, evenly. It depends on the location of the elements of production, i.e., on geography, though not physical geography, but rather economic geography. To clarify:

In order for raw materials in the hands of a person to turn into craft products, fuel (or another source of energy: water, or wind) is necessary. The main thing is FUEL.

When the craft is separated from agriculture, such an upsurge of productive forces occurs, that if there is not enough local fuel reserves (which should be so), the fate of production must become dependent on the possibility of obtaining it from more remote areas. Thus, the lines of communication, not just as a means of exchange nor a means of trade, but as a condition for production and a means of delivering fuels (mainly forestry), become the FIRST source of the development of a small industrial, meaning the URBAN settlement (hence the exchange of goods).

Transport is the FIRST condition for placing industry, separating it from agriculture, and becomes the FIRST cause of crowding on the land. Because of the uneven distribution of fuel on the surface of the earth, with the ubiquity of raw materials due to the need for fuel to be transported to the raw materials, there is uneven distribution of industry, unevenness and dispersal of people following the placement of industry, and unevenness in the size of cities.

The centers of attraction of fuel became the centers of gravity of raw materials. The means of distribution fuel became also the means of distribution of raw materials (hence trade).

Thus, transport is reinforced as the reason for concentration of fuel, raw materials, and housing.

Indeed, what is the city? It is a node of means of transportation: a node of roads and rivers.

Why is the city different from the village? There are intersections of roads, streets, and rivers. The village, unlike the city, always has just ONE ROAD. If there is more than one road in the village, surely there has been a qualitative leap in its inner structure, and the craft has been separated from agriculture. It can still be called a village, not a city, but it cannot be rural anymore. This is the BEGINNING of the city, and the END of the village, its DEMISE.

Still, what causes the gravitation of people to one point? What causes people to pile their dwellings on each other? Time. Socially necessary time. Let us not forget that distance is measured by time.

Production is developed better in quantity, in quality, and in pace where there is the easiest, fastest, and most convenient way to receive fuel (and later raw materials and then semi-finished products) in order to be able to make products in the same given time but easier, faster, and better. The craftsman is always (as well as the peasant) in pursuit of time. Time is a product, time is a commodity. That is why, it is necessary for him to put housing on the crossroads of the ways delivering to him fuel (and later raw materials and semi-finished products).

Fuel! Transport of fuel! The revolution in the transfer of energy is the condition that allows us to hope for the removal and incredible isolation on the one hand and the extraordinary, hypertrophied crowding on the other.

Energy can be transferred to a distance by wire. The wire began to replace the road. Transportation of fuel has turned into TRANSPORTATION OF ENERGY.

It must be borne in mind that the longer the wire, the more expensive the energy is. Therefore, for the first time, imported coal is converted into energy in old industrial centers, and then the industry passes to new, local sources of energy because they do not require transportation of fuel. This causes some zahirenie (withering) of former centers, but creates not smaller and sometimes even larger industrial centers, large cities although in new areas.

Only the spread of the NETWORK of wires to very remote corners of the country, only the creation of local stations and especially the so-called circular transmission of energy (a common network of various stations) creates the conditions for a complete revolution in the forms of accommodation and resettlement, a revolution as human history has never seen.

The achievement is that if you do not have to transport fuel to Leningrad once the fuel in the form of electricity can be received at the place of receipt of raw materials, e.g., somewhere near the magnetic ore from the Ural Mountains, there is no need to transport ore to Leningrad. Along with the ore, waste also can be transported, all production waste and other material that will eventually go into the waste.

Moreover, in these conditions, the volume of freight reduces not only for fuel and raw materials, but also for finished products. Further, the transition to local raw materials, depending on the cost of transporting the finished product, will help to abolish the narrow industrial SPECIALIZATION of the districts and their agricultural MONOCULTURE.

All modern development of agriculture is going in the direction of reducing cargo transportation of a product from the place of manufacturing to the place of its processing and consumption. The selection and the whole technique of plant hybridization, the fertilizer technique, the translucence of seeds, the greenhouse technique, and all the work of modern zootechnics are all focused on one task: PRODUCE ANY PRODUCT IN ANY GEOGRAPHICAL REGION.

The consequences of changes in the way energy is transported, the consequences of replacing the fuel paths with metal wires, especially the circular transmission, and wireless transmission in the future cannot be calculated and taken into account carelessly.

Raw materials were developed near the fuel, and raw materials were attracted to the fuel, to the crossroads of communications.

Now the energy reaches for the raw materials. Raw materials are everywhere—energy is stretched everywhere. Production is everywhere—housing is everywhere. The arrival of energy to local raw materials radically changes the position of agriculture. Agriculture is distributed not by centers, but across the entire land plot. How did this continuity of the agricultural production predetermine the great development of urban city centers?

The village, as the main engine of development of the city and a source of raw materials for the city (that is a commercial, not a communal village), is developed based on its links with the market and the city, i.e., it stretched out with its raw materials to the fuel, i.e., to the city, to the intersection of the flows of raw materials with the flows of fuel. The fuel was located on the intersection of flows (but not necessarily where it was mined). So were raw materials. Now energy is drawn to agricultural raw materials, NOT TO THE INTERSECTION OF FLOWS, but TO THE PLACE OF MANUFACTURING OF THESE RAW MATERIALS.

We reconstruct the village not only in the sense of its transition from small craft to large-scale, machine-made methods of production. More precisely, the transition of agriculture to the machine mode of production or the merging of industry with agriculture will mean that RAW MATERIALS WILL BE PROCESSED IN THE SAME PLACE as where it will be produced.

THE SMALLER THE ENTERPRISE for processing agricultural products, the shorter the distance from the agricultural raw materials to processing, and the shorter the distance to deliver the finished product that is not intended for consumption in other districts.

So the greatest economic concentration of production leads to the greatest spatial decentralization of production. Lenin called this process DISPERSAL of industry. In this way will industry merge with agriculture.

We do not know yet what this process will lead to, in the technical sense. But already now it is not difficult to imagine a COMBINE, which would perform, say one more additional operation—the transformation of grain into flour.

Thus, production would return from the enclosed premises to open air. The new conditions for energy transport pose the question of energy sources differently. Will the means of energy transmission—by wire or wireless—matter? Will energy be distributed from some enormous centers—energy sources? Will local energy sources also develop?

The network will replace the centers. In this network, energy centers are not of great importance, unlike the opportunity to assemble large, small, and the smallest sources of energy into one single power network. The transition from the centers of energy to the network turns the problem of the depletion of the world's energy reserves upside down, since the network deals with the collection of the smallest sources of energy, while the centers squander the largest ones.

How will these processes, as new forms of production location, affect the forms of settlement? How to put an end to both human isolation and housing overcrowding, these products of simple commodity society, infinitely amplified by large-scale capitalism? This is the question of:

RELATIONSHIP BETWEEN HOUSING CLUSTERS WITH PRODUCTION CLUSTERS.

Historically, the following four types of interaction between the clusters of dwellings in relation to the grouping of people in production are known.

First. COPRODUCTION CREATES COHABITATION. COPRODUCTION CREATES COMMUNAL HOUSING. This is confirmed by the example of primitive communism (shepherd cultures). In this culture, a dwelling was built for all participants of the production process. Researchers estimated 250–300 people in one dwelling.

Can this case be extended to other social formations? Communism, MODERN communism, should embrace at least hundreds of millions of people in a common production process. If joint labor led to cohabitation, it would be right to build one accommodation for several hundred million people. For a better verification of the situation, let us move on to the second question.

SEPARATE PRODUCTION LEADS TO COHABITATION.

The most independent, separate producer in history is the petty bourgeois. However, it is exactly the small-scale industry that created "a conglomeration of gigantic masses in cities". Take Paris, London, or Berlin. To guarantee accuracy, we will imagine them in the fourteenth–fifteenth centuries, when there was no large-scale industry.

If we believe that joint collective labor leads to collective living, to a common house, then it turns out that the petty bourgeoisie would have to reduce the existing villages in size, while in fact it turned them, these small villages, into condensed formations in the form of cities.

Third. SEPARATE PRODUCTION LEADS TO SEPARATE ACCOMMODA-TION.

This happens with the village under the influence of the development of commodity relations. In that way capitalism affects the village. That very capitalism which in the course of its development ostensibly creates overcrowding.

The second and third cases perfectly refute the assumption that the CAPITAL CONCENTRATION CREATES CONCENTRATION OF HOUSES. There is no capital yet, but the city is already there. There is no concentration of capital yet, but the city is already flourishing. Babylon itself would laugh at this idea.

But what about the village, which, contrary to these claims, is not crowded, but more than that, is being reduced in population due to the enlargement of production and concentration of capital?

Only the first and the third cases seem to confirm the dependence of cohabitation on joint labor and, consequently, separate living on separate work.

The fourth case. JOINT PRODUCTION LEADS TO SEPARATE LIVING.

The working barracks seem to confirm the accuracy of the law, derived from the experience of the first and third cases. We work together—we live together. Alas, this applies only to poor, underdeveloped capitalism, mainly in the era of manufacturing. In developed capitalism, the worker has a single family dwelling and even lives in a cottage whenever possible. We can condemn the "petty bourgeois inclination" of the worker, attack his backwardness, "opportunism", accuse him of biological bias, etc., but we do not know yet a single concrete worker who would exchange his apartment and let alone the cottage for a cot in a dormitory. But we know plenty of cases of the opposite.

Here, we turn to the question of what is the INTERNAL, SOCIAL FORM OF HOUSING.

Is the city a dwelling?—No. It is a form of settlement rather than a form of dwelling.

A house? A big house in the city—is it a dwelling? No. This is a vertical form of crowded dwellings, but not a dwelling.

A dwelling is a building that embraces the clan. A courtyard (but not a hut) is a dwelling. And the communal patriarchal village is no longer a dwelling; it is a form of resettlement.

A BARRACK or a DORMITORY is a dwelling.

A KHUTOR (hamlet) is a dwelling.

A FARM is a dwelling.

A COTTAGE, an APARTMENT, a ROOM, or a CORNER is a dwelling. But a city, a village, a house, or a hut is not a dwelling.

The communist dwelling does not recognize any divisions on the urban and rural: divisions of the village into yards, divisions of the city into houses, and divisions of the house into rooms.

This lack of divisions within the settlement and home signifies the absence of social division of labor. The patriarchal community already knows the first major division of labor in society: between a woman and a man. A man is engaged in agriculture, while the wife is engaged at home, hence her subordinate position in society. When there are no field works, the husband is at home engaged in craft, while the wife is again occupied with housekeeping. Why is one house of the clan, under these conditions, not divided into two halves—male and female, prior to the division of one dwelling into many, and then dwellings into yards?

The transition from cattle breeding to agriculture played a role in this. One dwelling transforms into a village consisting of many dwellings. What it depends on? On the transition to a new mode of production, meaning a new way of accommodation.

The rural dwelling is rather an economic than an arithmetical unit. There are as many male workers as necessary to cultivate a particular land plot. What is the size of this plot?

Its size is measured in time sufficient to enable the cultivation during the favorable weather with given tools.

The increase in the number of employees will lead to a fragmentation of the land plot according to the number of new workers and, consequently, the creation of a new dwelling.

Why? Because this production method requires certain cooperation in production, since every extra person becomes not only an employee helping production, but at the same time a consumer, a waster of the common product. The number of employees must therefore be minimal.

This number is what is called the patriarchal family, which replaced the communist clan. In this case, why are the dwellings located not separately, but next to each other in one village? Because the division of labor in agriculture is not ultimate. Part of the territory such as meadows, pastures, and forests remains in common use. Only the transition to a new mode of production and to a new division of labor leads to new forms of resettlement and to new forms of housing. Meadows, pastures, and forests unite people in villages (into one common grouping) as they did before, while farming splits them into separate yards (dwellings). But since the crafts become permanent and not seasonal occupation for part of the village, this new division of labor means the beginning of the city and the end of the village. This new division of labor also creates a new dwelling—a philistine dwelling, the home of an independent artisan. A new division of labor led to a new division and separation of the dwelling. The petty bourgeois family, unlike the patriarchal family, comprises only two generations, since the new division of labor obviously requires fewer people for the same operation.

The clan and the patriarchal family are replaced by a petty bourgeois family, then a capitalist family under capitalism in the city, but what about the village?

There is not only the separation of industry from agriculture, but the further division of labor in agriculture itself. Under capitalist agriculture, the producer is already detached from others, on the one hand, due to the strengthening of specialization, and on the other, by virtue of the separation of field crop cultivation from animal husbandry.

The community (hence the village) has disintegrated, and from the village, the person moves to the farm. What will happen under the socialist method of settlement?

There will be no city or village, there will be no boundaries between them, and they will unite as one.

And the dwelling? The dwelling is the product of a further technical division of labor, leading to further reduction in its size. After the clan dwelling, patriarchal dwelling spanning 4 generations, philistine dwelling for two generations, and capitalist dwelling for one, the technical division of labor under socialism will lead to a dwelling for one person—to an individual dwelling.

Why does this not happen under capitalism?

Because the wife and husband cannot yet make the final division of labor, as the capitalist is bound by the division of labor with the wage worker. They are connected by the unity of economic interests: inheritance of property and the unification of capital. Similarly, the proletarian family is bound by the common interests of the reproduction of its workforce and the hope of supporting its forces in old age by children.

Only the socialist relations will put society directly before the man-producer, and the man-producer will be directly faced only by public relations… For they destroy the division of labor between man and woman.

[It should be noted that the course of various forms of the family—patriarchal, philistine, petty bourgeois, capitalist and proletarian—is consciously provided in abstract formulas.

In life, deviations are inevitable. The patriarchal family (4 generations) can reach in size the genus—6 generations; philistine (2 generations)—the patriarchal (4 generations); capitalist and proletarian (1 generation, or a couple)—the philistine (2 generations)]. Form of settlement: the city, or the village, the form affects the EXTERNAL FORM OF THE HOUSING, but not the internal one.

It does not matter, whether I live in a separate barrack or in a dormitory, located in a big house. The social character of the form of housing has not changed. Either way, the form of dwelling is primitive-communist.

Similarly, it does not matter if I live in a separate apartment of a multi-family building, or my individual apartment is located outside the city and is therefore called a cottage. In both cases, the social form of the dwelling is same: family-philistine.

If I live in a separate room, it does not matter if it is located inside a huge house or not, because a room is a room, even if it was set up alone in the countryside, surrounded by greenery, silence, etc. Social character or rather a social form, the internal form of the dwelling has not changed.

This is the conclusion, if we break away from the obsessive physio-geographical criterion and will refer to socio-historical one…

In both cases, the fact of joint, collective production is obvious. But in the first case, it leads to a common dwelling, and in the second case to separate apartments for families and even to separate rooms per person.

With low technology, joint labor also meant a common place of habitation; with highly developed, capitalist technology (division of labor), joint labor creates a SEPARATE HOUSING, but (here and there) JOINT LOCATION, which is not the same.

Maybe, the commonality of location (house, quarter, city) is created by common-ality in production? No, not at all. That is the whole point. Work at Ford enterprises is a common work for one concentrated capital, but people live in different groups: some under Detroit, others in Ireland, the third in Constantinople, the fourth in Belgrade.

The same is true of any solid steamship business: some live on the shore, others on steamships. But all work in one enterprise, because if you stop the work of one of the groups, the other will stop.

The joint work of primitive peoples lead to joint housing. The joint work of modern peoples develops personality.

How, for what reasons does the personality develop in man? Due to the appearance of property. Property in its economic dimension, exploiting, overwhelming others, all other personalities except personalities—members of this exploiting collective, will disappear.

But the person will be born separately—not collectively. Eating, drinking, sleeping, dressing, etc., in one word CONSUMPTION will always be done separately.

In more developed society, individual consumption rights will increase. Let us not forget that socialism means the excess of products (the five-year plan is the aspiration to turn the shortage—the product of capitalism—into excess—that of socialism). Along with the private property, bourgeois and capitalist property, bourgeois and capitalistic personality will disappear; but personal property, personal consump-tion, personal initiative, personal level of development, personal hands, personal legs, personal head, brain will not disappear; moreover, these will be accessible to everyone, not just the lucky few, the "privileged" ones, as it was before socialism.

Personality represents the need for a socialist mode of production, originating from the depths of the capitalist mode.

Personality is the product of a separate (at the moment, territorially, but not economically separate) production position. Personality is the product of the tech-nical division of labor, not the social division. Not to recognize the collective, but at the same time to celebrate the PERSONALITY as Max Stirner does, means to admire the consequence, despising the cause. To praise the collective and ignore the personality means to praise the Russian language, but forbid using Russian words. This is what our modern Stirnerians are doing backward, the supporters of the specific shade of Proudhonist communalism.

You cannot oppose the individual to the collective and vice versa. The stronger the collective ties, the stronger the personalities comprising the collective, its compo-nents, and DISTINGUISHED BY IT. The stronger the personality, the stronger the collective to which it serves, and the stronger the class collective to which this personality serves...

Personality is not arithmetic, but a social unit.

In presocialist formations, the individual served as a unit in so far as it was an enslaving, exploiting, oppressive power. The others were not considered human. The task of socialism is not to perpetuate this attitude, but to change it.

It is necessary to destroy exploitation, to destroy its product—an aristocratic, capitalistic, philistine personality. But to destroy a personality in general would mean to destroy all division of labor between individuals. Our task is not at all to destroy the division of labor in general, as the ideals of the peasant community, represented by N.K. Mikhailovsky, demand.

Our task is to destroy the PUBLIC division of labor: the one existing between the entrepreneur and the wage worker, between a man and a woman, between the town and the country, and finally, between individual countries. For the evaluation of the division of labor between people, we consider the development of the productive forces of society as an inevitable, necessary, and indispensable condition. But at the same time, we do not differ at all with N.K. Mikhailovsky on the EVALUATION of the consequences of SPECIALISM, arising from the technical division of labor. With all determination, we condemn it. But while Mikhailovsky contraposes this specialism to the past with its low level of productive forces, we contrast it with the future, with specialization. Yes, even higher specialization. It is necessary to carry out the technical division of labor until there is an opportunity for an easy change of labor.

Then, finally, the social division of labor will be destroyed, the opposition between mental and physical labor, skilled and unskilled, responsible and simple, "women's" and "peasant", urban and rural activities, etc.

Only then a complete fusion of "labor" and "education" will become possible. While general education and physical education try to fill the failures of specialism, they try to be universal. With the division of labor between individuals, it is futile to fight with the product of this division of labor, personality; it is unnecessary and reactionary. Our historical task is only to eliminate the one-sidedness of the human personality, to transform the personality from the bourgeois limited with its professionalism, from capitalistically limited with its specialism to a comprehensively developed, socialist personality. We are for the division of labor, but we are also for a system of change of labor, which makes the division of labor possible.

The difficulty of mastering the essence of the relationship between the individual and the collective, due to the impossibility of solving it without studying the division of labor—the social and the technical division—is further aggravated by the fact that economic, industrial, and territorial scales are forgotten.

Without the ability to sharply distinguish between these three things, one cannot understand anything about the agrarian economy, as well as the problem of industrial dispersion. The economic size of an enterprise in modern society is measured by the size of capital, but not by the number of workers employed in the enterprise—otherwise, a huge power plant could be named a small workshop, since just a few people work at the electric power station; and another craft workshop of a manufactory type that has hundreds of workers would have to be declared a large factory. The production size is also determined not by the number of people involved in the tools of production. The production size of the enterprise is characterized by the number of products produced by the worker of equal preparedness, equal effort, with the same amount of time. Territorial size can sometimes coincide with economic and production scales.

Especially it is so in the era of transition from small to large-scale production, for example, from small urban craft to manufacturing or from small-scale farming to cooperative, collectivization.

Under certain conditions of location, centers became a sign of the largeness of enterprises; in new, different conditions, the network will take their place.

It is the network that will become the condition, the territorial sign of size.

What is larger, one laundry for 25 thousand inhabitants of the city, mechanized "according to the latest technology", or 25 laundries, each for a thousand people, also mechanized with an even latest technology? Of course, 25 laundries is greater than one, although it usually thought otherwise. They are more necessary, since most of the settlements are less than 25 thousand people. They will displace small laundries, for they are 25 times (in fact much more!) closer to the consumer. They will beat small equipment, because these laundry rooms are easy to produce in comparison with heavy equipment of large laundry ... But, of course, they are only possible when there is a network or at least local sources of energy.

Same with kitchen-factories, etc.

The world will someday come to such an enlargement that there will be only one network of food, washing, housing, etc. But a human will never see such an enlargement of enterprises that on the entire planet, there will be built just one laundry or dining room, one center for washing, or food.

The network will win, the center will die out.

In the future, the crowdedness of people in production will be unthinkable. On the contrary: around one group of automatic machines, which employs one person, it will be possible to leave free space for greenery, light and air, and production will cease to be a source of occupational diseases, connected largely with the location of the enterprise. Such is the perspective of the inevitable result of the gradual displacement of man by the machine.

Reduction of the time required for production of the product caused overcrowding in these conditions (of obtaining energy and raw materials), creating an unprecedentedly high land rent. This same reduction in time in other conditions, instead of congestion, concentration, clusters, will lead to the MAXIMUM SCATTERING of industry, to MAXIMALLY UNIFORM (i.e., without congestion on one side and deserts on the other, one that evokes the existence of the other...) resettlement of people.

It is interesting to compare our views with the views of the anarchist communists in the person of their ideologist P. Kropotkin.

According to Kropotkin, centers supply energy directly to the house of a local producer (in the future). In our view: there is no energy centers. There is a unified NETWORK of energy. There are no central sources; there are local, large, small, and smallest, ubiquitous sources of energy. Each center is a periphery, and each point of the periphery is a center.

The network should be a collector of the smallest energies and not just a way of distributing energy from large sources.

P. Kropotkin frees a human from crowding by returning to small craft with the help of large-scale industry.

Scientific communism fights crowding and isolation through the new placement of agriculture and mining, by combining the manufacturing industry with them into one.

TERRITORIAL solutions are close, almost identical: both Kropotkin and Lenin have the SCATTERING of production. But economically, one stands on the basis of small craft, the other—a large-scale mode of production.

And these former conditions for the location of production and settlement of people affected the very METHOD OF BUILDING PRODUCTION. Everyone talks about the backwardness of the construction industry. We must not talk, and we must explain this backwardness.

...Capitalism is developing unevenly. Next to the greatest, sophisticated modern large-scale industry, there are infinitely backward:

(1) AGRICULTURE
(2) construction technology
(3) techniques of food processing, washing clothes, etc. (so-called housekeeping).

These three backward industries have not yet reached the level of other industries.

The main reason for the backwardness of construction equipment and agriculture is the same.

The decisive shift in the technology of agricultural production is associated with the appearance of gasoline and technical transport in the countryside. For the first time, energy came to raw materials and construction materials came to village. For the first time, energy came to raw materials. As for the construction technology...

The modern direction of its development is connected with the same movement of energy to any local raw materials. The transfer of energy over increasingly long distances will end urban land rent (here the technical revolution will help the social); this last concrete, durable and heavy construction is an obstacle to revolution in construction technology.

In fact, why in the most developed centers of modern industry, where there is both raw materials and fuel, where energy was concentrated, why, in these seemingly most favorable conditions, construction production remained at the handicraft-medieval level?

Because of the high land rent. Because of the gravitation of industry to the crossroads of fuel and raw materials, semi-finished products.

External crowding led to the internal crowding. Everyone's rush to the place of production reduced the area of each place, and any expansion of the dwelling could occur only at the expense of a neighbor. Thus, the technical ideal of the builder of the city became a dwelling of minimal dimensions.

Houses turned into boxes, receiving the shape of boxes. For the external form of buildings is changing depending not on the need or the internal social form of the dwellings and not depending on the building material (clay, wood, stone were and are used to build round and rectangular houses), but because of the DEPENDENCE OF THE EXTERNAL FORM ON THE LOCATION OF THE DWELLING, ON THE GROUPING OF DWELLINGS.

You cannot make a vaulted ceiling, for above you there is another dwelling, another room. And the need for residents (or for you) to walk on a flat floor inevitably makes you, living under them (or under yourself), having an even ceiling. You cannot change the shape of the dwelling in a horizontal direction, nor increase its size, for the external form of your home; the size of your home is limited by the arrangement, grouping of dwellings.

The road and the fence are the first reason for moving the dwelling to a "rectangular" shape from a "round" one. The ax made this transition from closed curves to straight lines of houses possible. Streets intersections and vertical grouping of dwellings are the reasons for the further development of rectangular houses.

The durability of buildings was the need for a manufactory mode of production and the inevitable consequence of crowding (in particular, vertical).

Changes in the technology of production and the need for the TECHNICAL RE-EQUIPMENT OF ENTERPRISES (pace, pace!) demand the destruction of old cities, old houses. They require the most short-lived buildings. Number of floors—this scourge of fragility—the last guard of the durability of buildings, must give way to floorless buildings.

A new way of production allocation has become the reason (which is already in effect today) that mankind would finally be able to move to a new method of production in the construction industry. Like every new revolutionary mode of production, it will come as the CHEAPEST...

The new way of building production will put an end to byt (the way of life), to the way of life in general. For construction should not perpetuate the relations of people (and byt, way of life is characterized by the permanence of relations), new construction should not displace them either. It must FOLLOW the person, the changes in the relations between people, and the development of the individual person. FOLLOW him, DO NOT overtake them, and do not consolidate the old relationship.

Dwelling, like clothes, can be improved: by increasing it in size, in breadth, in height, increasing the size of windows, etc. Is this conceivable under the old technology?

No, only prefabricated houses that can be assembled, disassembled, reassembled will meet the actual needs of a particular constantly developing person.

While a new way of production allocation requires a NEW GROUPING OF HOUSING, a new way of building production makes this task feasible.

But as long as we have not yet destroyed the existing real city, until the urban industry is SCATTERED, we should consider about 40–50 years, in which we must lay down our projects of resettling the workers of these old cities. Buildings are not depreciated until the expiration of this (approximate) period, for energy is still obtained from the centers, not from the network. In this era, passenger transport will play a huge role—especially the car. The more the city develops, the greater is the need for intra-urban transport. The more the city develops, the smaller is the opportunity of using this transport. This is the result of crowding, accumulation, concentration. This is the manifestation of the contradiction between the urban form of allocation and the producer's desire for production.

The contradiction is so great that sometimes it is better for the modern worker to live further (geographically) in order to be closer (in time, although this is possible only under the condition of USING THE PASSENGER TRANSPORT as a means of transportation to the place of production).

Thus, the passenger VEHICLE REPRESENTS THE NEED OF A TRANSITIONAL EPOCH from the city to its destruction and uniting with the village into one; the car is an inevitable companion of the spatially unscattered yet industry.

In the transitional era, any form of housing—whether the philistine one, the household commune, the proletarian or, finally, the socialist, individual dwelling— all of them will inevitably imprint coercion and closeness and remoteness, since the HOUSING MATERIAL is motionless, unchangeable for the continuation of the ENTIRE CENTURY, while the form of the dwelling is changed within half a decade, and sometimes even half a year.

In this way, the method of building production on the one hand and the planning method on the other becomes a means of resolving social issues.

They will either interfere (forcedness is exactly in this) or facilitate the formation of a new social form of housing and will prevent new relationships between people or stimulate them. Under these conditions, the human is left with the choice to force the technology to serve society, serve the people, their needs; a man under these conditions will have to declare war, a relentless war on technology, old construction technology, and its ideological representative—unprincipled imitation, boundless technicism. On the old technology, he will declare war. New machinery, he will harness in the chariot of history.

Printed by Printforce, United Kingdom